VLSI Circuit Design
Methodology Demystified

Use this book to identify where
MCU has needs, and develop yourself
in those areas.

VLSI Circuit Design
Methodology Demystified
A Conceptual Taxonomy

Liming Xiu

IEEE PRESS

WILEY-INTERSCIENCE
A JOHN WILEY & SONS, INC., PUBLICATION

Library of Congress Cataloging-in-Publication Data is available.

ISBN 978-0-470-12742-1

Printed in the United States of America.

10 9 8 7 6 5 4 3 2 1

To my wife, Zhihong,
my daughters, Katherine and Helen,
and my parents, Zunxin and Zhengfeng

Contents

Foreword

Richard Templeton
President and CEO
Texas Instruments, Inc.

Few inventions have had greater impact on the everyday lives of people around the world than the integrated circuit. I often wonder if Jack Kilby had any inkling of what was ahead for himself and Texas Instruments when he sketched out that first integrated circuit in his engineering notebook in 1958.

Yet, even as the term "computer chip" has entered the vocabulary of millions, the knowledge of how modern integrated circuits are actually designed remains the provenance of a relative few. Even among college-educated electrical engineers with a deep understanding of physics, electrical theory, and logic, the broad concepts behind designing a modern IC comprised of hundreds of millions of transistors can be unexplored territory.

Add to this knowledge gap the fact that after a period of relatively predictable scaling of transistors and computational performance, the industry is starting to see fundamental limits of the standard materials and techniques used in building these amazing products. Power consumption, heat, and cost of modern system-on-chip products may turn out to be the ultimate tests of how far we can take microelectronics, not how small we can make individual transistors. To solve those problems, we need to arm our students, teachers, and working professionals with a solid, fundamental understanding of how modern chips are made and the engineering challenges associated with that work.

Why? Because the rewards we see in continuing the advancement of the science of chip design are too great to ignore. Take health care as one example. As system-on-chip design makes smaller, portable, and affordable diagnostic and treatment products possible, health care becomes more personal. Individuals will have quicker access to information and treatment, and affordable electronics can put that capability in the hands of a much wider range of people.

Wireless communications will be an essential part of this shift, but not just in medical electronics. Historically, most analog and RF electronics have been implemented in technologies several process generations behind state-of-the-art digital system-on-chip products. Hence, a substantial opportunity exists to enhance system performance, cost, and power by moving RF processing into the digital domain. This is being done today, but the opportunity to connect not just people, but billions of devices to each other and the Internet is just now starting to be realized and is fertile ground for the innovative chip designer.

As a senior member of the Technical Staff at Texas Instruments, Liming Xiu is closely connected to the challenges associated with chip design on a daily basis. As an experienced educator and general chair of the IEEE Circuit and System Society, Dallas chapter, he has demonstrated a passion for sharing what he knows with others so they can advance their own capabilities. I congratulate him on successfully tying together a wide range of highly technical topics into this comprehensive, insightful overview.

Foreword

Dr. Hans Stork
Senior Vice President and Chief Technology Officer
Texas Instruments, Inc.

It takes many years of study and experience to acquire the breadth of knowledge required to participate successfully in the design of complex systems on silicon. There are many courses taught at universities or perhaps online that one can take to learn about topics such as silicon process technology, device physics and fabrication, circuit design, logic design and verification, high-level system synthesis and description languages, and so on. Typically, these are stand-alone, detailed, mostly theoretical foundations for specific areas of study. However, when it comes to playing an effective role in a product design team, it is critical to have some insight into the practical aspects and to have at least a limited understanding of what your team members are doing. Since none of us has all the knowledge at hand in all situations, we typically wind up asking questions of our colleagues. That is after we have wasted several hours trying to find the information online or in one of our textbooks, which we thought we studied so well, perhaps even after overcoming any anxieties about looking dumb or being apologetic for taking time away from others. Wouldn't it be great to have access to a book that anticipated our questions and had first-rate answers?

Imagine my pleasant surprise when Liming Xiu asked me to write a foreword to his book, which was, in fact, his taxonomy of questions and answers for people engaged in complex VLSI product design. An experienced designer and integration engineer himself, he recognized the need for a framework that provides answers to many questions for active designers of silicon systems on a chip. His personal experiences were the inspiration to write this up-to-date, comprehensive summary of knowledge. You will find some of the most succinct descriptions of topics as varied as device operation, RTL description, and functional and test coverage, as well as solid yet simple explanations of why power is a growing concern in sub 100 nm CMOS and of why LVS is so difficult even with the most advanced EDA tools. Liming does a remarkable job of getting to the heart of the issues and provides an extensive bibliography for follow-up.

Frankly, I have always been somewhat skeptical that a book could strike a balance between being practical and being fundamental. This one, written in the creative way of answering typical workplace questions, goes a long way toward meeting that need. I think many novice as well as experienced system architects, but also designers and process definition and integration engineers, will like the simple access to a preliminary answer that this book provides. The logical outline and the comprehensive index and glossary make the information easily accessible. It will build readers' breadth of knowledge, which is so important to any good integration engineer.

Preface

The widespread acceptance of sophisticated electronic devices in our daily life and the growing challenges of a more technically oriented future have created an unprecedented demand for very large scale integration, or VLSI, circuits. Meeting this demand requires advances in material science and processing equipment and the ability to use the computer more effectively to aid the design process. More importantly, it requires a significant number of qualified and talented individuals to work this ultra-complicated task. The goal of this book is to equip interested individuals with the essential knowledge to enter this exciting field and to help those already involved to reach higher levels of proficiency.

The challenges in VLSI chip development come from several directions. Market pressure constantly demands shortening of the product development cycle. The maximization of profit margin requires control of engineering cost. Furthermore, the product not only has to be successful technically, but also financially. These challenges put tremendous strain on the execution of the VLSI chip development process.

The art of VLSI circuit design is dynamic; it evolves constantly with advances in process technology and innovations in the electronic design automation (EDA) industry. This is especially true in the area of system-on-chip (SoC) integration. Due to its complexity and dynamic nature, the topic of VLSI circuit design methodology is not widely taught in universities, nor is it well understood by many engineers in this industry. The objective of this book is to give the reader the opportunity to see the whole picture of how a complex chip is developed, from concept to silicon.

This book primarily addresses the group of people whose main interest is chip integration. The focus of chip integration is implementation, not the circuit design itself. Unlike transistor-level circuit designers who spend most of their time on the architecture, analysis, optimization, and simulation of small circuit components, chip integration engineers (or implementation engineers) mostly work on the task of turning a large chip design from a logic entity (RTL description or netlist) into a physical entity. The spirit embedded in this activity is "put everything together and make it work," not "create/invent something from scratch." Consequently, working as an IC implementation engineer requires a unique set of knowledge and skills.

This book has grown out of lecture notes prepared for graduate-level students. A technical background in introductory-level circuit design courses and introductory digital logic courses is required to understand its contents. Due to the dynamic nature of VLSI design methodology, this book is not organized by chapters; rather, it is organized in a questions-and-answers format. Further it is organized in the order of chip development: logic design, logic verification, logic synthesis, place and route, and physical verification. By demonstrating the key concepts involved in VLSI chip development process, it is my hope to help readers build a solid foundation for further advancement in this field.

LIMING XIU

Dallas, Texas
July 2007

Acknowledgments

During the development of this book, I received invaluable help from many people, especially from my colleagues in Texas Instruments, Inc. I have not named these individuals in the text since the list would be very long, however their help is greatly appreciated.

A number of reviewers were instrumental in improving the quality of the manuscript. Their detailed comments on the first draft were very helpful. I thank them as well. I also thank Mary Miller of Micron for technical writing assistance.

I have listed all the work other than my own in the references as much as possible. However, there are still some materials that I cannot credit to the sources, such as those obtained from the Internet, product datasheets, non-technical magazines, and so on.

L. X.

CHAPTER 1

The Big Picture

1. WHAT IS A CHIP?

A *chip* (*integrated circuit* or *IC*) is a miniaturized electronic circuit that is manufactured on the surface of a thin substrate of semiconductor material.

Functionally, a chip is a hardware component that can perform certain desired functions. For example, a simple chip might be designed to perform a simple function of a logic NOR (such as the 4000-series CMOS, dual 3-input NOR gate and NOT gate shown in Figure 1.1), a simple operational amplifier, or an *analog-to-digital converter (ADC)*. However, a complex *system on chip (SoC)* performs much more complicated tasks (see Figure 1.2). Examples include those for video decoders, cellular phones, network routers, or general-purpose CPUs for personal computers.

Structurally, chips are manufactured on a semiconductor material called silicon. Basic components such as transistors, diodes, resistors, inductors, and capacitors are constructed on the silicon. Those basic components make up the chip, simple or complex. Simple chips may only contain hundreds of those basic components, whereas complex chips may contain hundreds of millions of those components. Since 1959 (the year that the first integrated circuit-related patent was filed by Jack Kilby), several terms has been created to reflect the status of integrated-circuit development: *small-scale integration (SSI)* for tens of transistors on a chip, *medium-scale integration (MSI)* for hundreds of transistors per chip, *large-scale integration (LSI)* with tens of thousands of transistors per chip, and *very large scale integration (VLSI)* with hundreds of thousands of transistors. *Ultra large scale integration (ULSI)* and system-on-chip (SoC) are the latest terms to cover the modern, ultracomplicated chips with billions of transistors on a single chip. All chips are roughly classified as one of three types: purely digital, analog, or mixed-signal.

Application-wise, chips can be designed to target various applications: video/graphic, audio, communications, networking, general-purpose personal computing, supercomputing, automotive, industry control, medical instrument, and military.

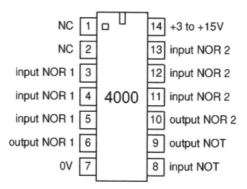

Figure 1.1. 4000-series CMOS, dual 3-input NOR gate and NOT gate.

The majority of the today's chips are designed for processing signals or manipulating information. Among the tasks performed are collecting, transporting, presenting, processing, or manipulating all kinds of information. And today, information plays a vital role in our daily lives. There are billions of billions of bits of information generated every day to support the normal operations of human society. Every single one of those bits must be processed by some kind of chip. Thus, it is not a surprise that the semicon-

Figure 1.2. System-on-chip example.

ductor chip is built into our life. In addition to this information-processing chip, there are other types of chips that interface with our activities directly by driving electrical, mechanical, or optical components that result in something that we can see, hear, feel, or smell.

Finally, an indivisible part of the chip is the associated software. Software enables the chip to perform certain specific tasks. Software tells the chip *when* and how to do *what*. Without software, the chip is useless, just like a human without brain. Well-developed software can perfect the chip's feature sets, can prolong the chip's life, and can make the difference between success and failure.

When used, a chip is packed in a package, which is mounted on a *printed circuit board (PCB)* and installed in an end-equipment system.

In summary, a chip is an entity that has a large number of transistors integrated into it. Constructing circuits in this manner is an enormous improvement over the manual assembly of circuits using discrete electronic components. Two primary advantages are cost and performance. Cost is low because the components within a chip are built as one unit and not constructed one transistor at a time. Performance is high because the integrated transistors switch quicker and consume less power due to the fact that the components are small and close together.

2. WHAT ARE THE REQUIREMENTS OF A SUCCESSFUL CHIP DESIGN?

In the field of modern VLSI circuit design, constructing a chip from concept to silicon is an ultracomplicated task that involves many factors. For a successful project, the chip must be:

- Structurally correct to achieve its intended design functions
- Functionally correct at the designed clock speed in various working environments (voltage, temperature, and process corner)
- Reliable throughout its life (e.g., 100k hours or eleven years)
- Manufacturing-friendly

Further, it must be built such that:

- It can be handled safely in an assembly line and various other environments without being damaged (e.g., it is protected from *electrostatic discharge* or *ESD* and *latch-up*).
- It can be packaged economically.

- It stays within its power budget.
- Cost is minimized.
- It is manufactured within its time schedule.

And, then, finally, there must be an existing or potential market for this chip.

3. WHAT ARE THE CHALLENGES IN TODAY'S VERY DEEP SUBMICRON (VDSM), MULTIMILLION GATE DESIGNS?

Designing a system-on-chip (tens of millions of gates and larger) in a very deep submicron (90 nm and below) environment is a task of solving many complicated, interdependent problems at once. The design/implementation/verification methodology required is a dynamic development since the challenges involved are ever-changing as the process technology continuously advances. The most outstanding challenges today are listed below:

- *Timing closure.* Timing closure is often the most difficult task in designing a chip owing to the fact that a logic gate's timing behavior (or speed) varies greatly at different temperatures, supply voltages, and process conditions under which the device is built and operated. Moreover, a logic gate's speed is also affected by the drive and load environment surrounding the logic gate. Timing closure means that the chip must run at a designed speed (represented by clock frequency) reliably under all conditions. This is not an easy task to achieve, especially when the process shrinks to even finer geometries and wire delays become more dominating in the overall delay equation.
- *Design verification.* Modern SoC devices contain a large number of components on board, such as processors, memories, on-chip busses, special function macros, and so on. The task of design verification is to ensure that the components work together faultlessly as designed. The magnitude of difficulty involved in this task increases dramatically as integration levels continuously grow and design sizes correspondingly increase.
- *Design integrity.* Design integrity includes *cross talk, IR drop, electromigration (EM), gate oxide integrity (GOI), electrostatic discharge (ESD),* and *latch-up protection.* The chip must be free of these problems before delivery for field application. These issues will become increasingly difficult to resolve as process technology advances.

- *Design for Testability.* The design must be testable for production defects. This testability must be built into the chip. As process geometry continually shrinks, new defect mechanisms constantly surface. As a result, design for testability is a subject investigated unceasingly by process scientists, design engineers, and tool developers.
- *Power budgeting and management.* Modern SoC chips can support more functions and perform tasks at higher speeds. Consequently, they tend to use much more power. In consideration of chip packaging, heat dissipation, and battery life, the chip's power consumption must be reduced.
- *Packaging.* As chips bear more I/Os and consume more power and I/O signals travel at higher speeds, chip packaging becomes more challenging.
- *Design reuse.* Characteristic of the SoC approach is the integration of components, rather than the design of individual components. The more components that can be reused from previous projects, or from other sources, the lower the development costs and the faster the project execution pace.
- *Hardware/software codesign.* Traditionally, software development cannot start until the hardware (the chip) is available. A new methodology, or design environment, is needed to solve this problem.
- *Clock management and distribution.* As a chip's clock speed increases and its clock structure becomes more complex, clock-related design issues will become more challenging.
- *Leakage current management and control.* As process geometries shrink below 90 nm, device leakage current increases dramatically. This problem has moved from backstage to front stage.
- *Design for manufacturability.* As process geometries shrink, device manufacturing requires more rigorous control. This fact imposes additional constraints on the chip design process.

4. WHAT MAJOR PROCESS TECHNOLOGIES ARE USED IN TODAY'S DESIGN ENVIRONMENT?

The mainstream process technology used in today's chip design/manufacturing environment is complementary-symmetry metal oxide semiconductor (CMOS) technology. Other technologies include bipolar, biCMOS, silicon on insulator (SOI), and gallium arsenide (GaAs).

In CMOS technology, *complementary symmetry* refers to the fact that a

CMOS circuit uses symmetrical pairs of p-type and n-type MOSFET transistors for logic functions. Originally, the phrase *metal oxide semiconductor* was a reference to the metal gate electrode placed on top of an oxide insulator. However, in today's CMOS processes, instead of metal, the gate electrode is comprised of a different material, polysilicon. Nevertheless, the name CMOS remains in use for the modern descendants of the original process. Today, in terms of dollar amount, the majority of integrated circuits manufactured are CMOS circuits. This is due to three characteristics of CMOS devices: high noise immunity, low static power, and high density.

The CMOS process has consistently advanced to smaller feature sizes over the years, allowing more circuitry to be packed in one chip, as described by Moore's law. This is the empirical observation made in 1965 by Gordon E. Moore (cofounder of Intel Corporation) that the number of transistors on an integrated circuit for minimum component cost doubles approximately every 24 months. Although Moore's law was initially made in the form of an observation and forecast, it has gradually served as a goal for the entire semiconductor industry. During the past several decades, it has driven semiconductor manufacturers to invest enormous resources for specified increases in processing power that were presumed to be soon attained by one or more of their competitors. In this regard, it can be viewed as a self-fulfilling prophecy.

As feature sizes shrink, costs per unit decrease, circuit speeds increase, and power consumption drops. Therefore, there is fierce competition among the manufacturers to use finer geometries. The status of current processes and the anticipated progress over the next few years is described and documented by the *International Technology Roadmap for Semiconductors (ITRS)*. Currently (in 2006), process geometries have dropped well below one micron. Today, 90 nm technology has been widely used for commercial products. In the near future, it is believed that the 65 nm technology will move front stage. And, tomorrow, 45 nm and 32 nm technologies will take the lead.

Copper has replaced aluminum for wire interconnect signal propagation material due to its improved electric conductivity. The interconnecting metal level has also been increased from two to six or even to seven or more. The power supply voltage for semiconductor chips has continually dropped, due to the shrinking of transistor size, to the current level of 1.1 V.

Table 1.1 presents typical data for each CMOS technology node, in which L_{drawn} represents the minimum transistor channel length and V_{DD} is the supply voltage for transistors. Lower V_{DD} can reduce the transistors' power usage. Density measures the number of logic gates that can be packed into one square millimeter of silicon. Unit gate capacitive indicates

Table 1.1. Typical metrics for CMOS technologies

	180 nm	130 nm	90 nm	65 nm	45 nm
L_{drawn} (nm)	180	95	60	50	40
Metal Level	4–5	5–6	5–6	6–7	7–8
Density (kgates/mm^2)	~ 70	~ 140	~ 250	~ 650	~ 1200
V_{DD} (V)	1.8	1.5	1.2	1.2	1.1
V_T (V)	~ 0.5	~ 0.45	~ 0.45	~ 0.4	~ 0.35
Unit gate capacitive (fp/μm^2)	~ 8	~ 10	~ 10	~ 10	~ 10
Metal resistance (ohms/square)	~ 0.1	~ 0.1	~ 0.12	~ 0.2	~ 0.3
Minimum metal width (μm)	0.25	0.175	0.13	0.1	0.07
Metal pitch (μm)	0.5	0.35	0.27	0.21	0.14

the capacitive loading of transistors, which has a great impact on the logic gate's speed. Metal level is the number of metal layers used for interconnecting. When more metal layers are used, higher densities can be achieved (but the cost is greater). Metal resistance measures the quality of the metal as interconnect material. Minimum metal width and metal pitch indicate the minimum width of metal that is allowed in a chip layout and how close metal can be placed to metal. These two parameters, together with L_{drawn}, primarily determine the gate density of the technology. V_T is the threshold voltage that controls when the NMOS or PMOS transistor switches. The level of V_T has great impact on noise margin and leakage current.

Bipolar refers to an electric circuit made of bipolar junction transistors, which were the devices of choice in the design of discrete and integrated circuits before the 1980s. It offers high speed, high gain, and low output impedance. However, its use has declined in favor of CMOS technology due to its high power consumption and large size.

BiCMOS is a technology that integrates bipolar and CMOS together to take advantage of the high input impedance of CMOS and the low output impedance and high gain of bipolar. A typical example of a BiCMOS circuit is a two-stage amplifier, which uses MOS transistors in the first stage and bipolar transistors in the second. However, BiCMOS as a fabrication process is not nearly as mature as either Bipolar or CMOS. It is very difficult to fine-tune both the bipolar and MOS components without adding extra fabrication steps and, consequently, increasing the cost.

Silicon on insulator (SOI) is a layered structure consisting of a thin layer of silicon fabricated on an insulating substrate. This process reduces the amount of electrical charge that a transistor must move during a switching operation and thus increases circuit speed and reduces switching energy (an improvement over CMOS). Moreover, SOI devices are inherently latch-up resistant, and there is a significant reduction in transistor leakage current, which makes this technology an attractive choice for low-power circuit design. However, the production of SOI chips requires restructured CMOS fabrication methods and facilities. Thus, it costs more to produce SOI chips, so they are generally used for high-end applications.

As we move toward 45 nm and 32 nm nodes, *multigate FETS (MuGFETs)* are increasingly being considered as a necessary alternative to keep pace with Moore's law. MuGFET is the general term for the class of devices that gain extra component width by allowing vertical active gates. FinFETs and trigates are examples of these devices. This new technology relies heavily on using high-quality, very thin SOI wafers as a starting material. Another popular SOI technology is *silicon on sapphire (SOS)*, which is used for special radiation-hardening applications in the military and aerospace industries.

Gallium arsenide (GaAs) is a semiconductor that has some electrical properties that are superior to silicon's. It has higher saturated electron velocity and higher electron mobility, allowing it to function at much higher frequencies. GaAs devices generate less noise than silicon devices. Also, they can be operated at higher power levels than the equivalent silicon device because they have higher breakdown voltages. These properties make GaAs circuitry ideal for mobile phones, satellite communications, microwave point-to-point links, and radar systems. However, high fabrication costs and high power consumption have made GaAs circuits unable to compete with silicon CMOS circuits in most applications.

5. WHAT ARE THE GOALS OF NEW CHIP DESIGN?

When a company makes a decision to invest in a project to create a product (designing a chip), the ultimate goal is to generate maximum profit from this investment. The approach to pursuing this goal is by conducting business "faster, better, and cheaper."

Faster means that the new chip must operate faster than its predecessors or faster than similar chips produced by competitors, which requires it to perform specific tasks in less time.

Better refers to the fact that the chip must support more functions (do more) than its predecessors.

Cheaper means that the cost of developing and manufacturing the new chip must be kept to a minimum.

This desire to develop something "faster, better, and cheaper" has motivated scientists and engineers working in this field to make enormous technical strides and will continue to drive them as they work to create superior products. This will, in turn, make our lives more enjoyable.

Perhaps the only exceptions to this ruthless pursuit are projects that are research oriented or not for profit or those produced for the government. In these cases, cheaper is not a concern. Therefore, faster and better can be pursued at whatever cost is required and on whatever schedule is demanded.

6. WHAT ARE THE MAJOR APPROACHES OF TODAY'S VERY LARGE SCALE INTEGRATION (VLSI) CIRCUIT DESIGN PRACTICES?

The major approaches for modern chip design practice follow:

- Custom design
- Field programmable gate array (FPGA)
- Standard cell-based design (ASIC)
- Platform/structured ASIC

In the *custom design* approach, each individual transistor is designed and laid out manually. The main advantage of this method is that the circuit is highly optimized for speed, area, or power. This design style is only suitable for very high performance circuitries, however, due to amount of manual work involved.

Field programmable gate arrays (FPGAs) are semiconductor devices that are comprised of programmable logic components and programmable interconnects. The programmable logic components are programmed to duplicate the functionality of basic logic gates, such as AND, OR, XOR, or NOT gates, or more complex combinational functions, such as decoders, or certain simple math functions. Structurally, the FPGA approach is a chip implementation methodology in which the base layers are premanufactured. When implemented in FPGA, only metal layers need be programmed.

In the past, this design approach was reserved primarily for emulations and prototypes. However, there are increasingly more FPGA-based products surfacing as this method gradually becomes mature and efficient. FPGA could offer an attractive alternative for low-volume commercial products because it has a lower *nonrecurring engineering (NRE)* cost. It also has a shorter time to market and can be reprogrammed in the field to fix bugs and so on. However, compared to ASIC, its unit cost can be much higher. Thus, for large-volume products, the ASIC approach is the better choice. Furthermore, due to its structure, FPGA performance is often inferior to that of ASIC.

Standard cell methodology is a method of designing *application-specific integrated circuits (ASICs)* with mostly digital content. A standard cell is group of transistor and interconnect structures that provides a Boolean logic function (e.g., AND, OR, XOR, XNOR, or inverter) or a storage function (flip-flop or latch). The cell's Boolean logic function is called its *logical view*. Its functional behavior is captured in the form of a truth table or Boolean algebraic equation (for combinational logic) or a state transition table (for sequential logic). From a manufacturing perspective, the layout of the standard cell (an abstract drawing of polygons) is the critical view. Layout is organized into base layers, which correspond to the structures of the transistor devices, and interconnect layers (metal layers), which join the terminals of the transistor formations. In design practice, the layout view is the lowest level of design abstraction.

The invention of logic synthesis and place and route tools has enabled the standard cell design style or ASIC approach. In this approach, the standard cells and other preassembled macro cells are grouped together to form an ASIC library. The chip functions are achieved by the cells in the library and through logic synthesis and physical place and route. In this approach, as contrasted to FPGA and platform ASIC, a mask is required for every layer, including the base and metal layers. The NRE cost associated with ASIC is often high due to the design, verification, implementation, and mask cost. However, for very large volume, high NRE costs could be offset by relatively low manufacturing costs.

Standard-cell ASIC methodology together with semiconductor process advances are the two major factors that have enabled ASIC chips to scale from simple, single-function ICs of several thousand gates to complex SoC devices of many million gates.

A platform-structured ASIC approach falls between an ASIC and a FPGA. It is an ASIC approach based on a preassembled *platform*. Inside the various platforms that a vender offers, certain special functions are already predesigned and verified. Random logic functions can be achieved by programming the metal layers in certain areas reserved for that purpose. Users can select a desired platform based on their needs. The main advantage of this platform-based ASIC is that the platform is already preassembled, which saves the mask cost of base layers. The only expense is the metal-programmable layers. Another benefit is that the verification costs can be significantly lower than those of an ASIC because the major functions on the platform might be preverified. As for performance, a platform ASIC often cannot match an ASIC; the design is not fully optimized as in the case of an ASIC. However, performance should be significantly better than that of an FPGA. In summary, the platform ASIC trades the high performance of an ASIC with shorter time to market and lower development cost.

The platform ASIC approach is gaining momentum due to its relatively lower NRE cost as compared to an ASIC. But for very large volume products, its unit cost could be higher than that of ASIC.

7. WHAT IS STANDARD CELL-BASED, APPLICATION-SPECIFIC INTEGRATED CIRCUIT (ASIC) DESIGN METHODOLOGY?

Standard cell methodology is a chip design approach that is based on pre-assembled library cells. The standard cells and macros, such as memories, I/Os, special-function cells, phase lock loops (PLLs), and so on, associated with this library are already designed, laid out, and verified in a predetermined process node. These cells are completely characterized and logical, timing, physical, and electrical models are already created and properly packed in the library. After a design is created in register transfer level (RTL) format, it can be mapped into those preassembled cells through a logic synthesis process by sophisticated synthesis algorithms. The resultant netlist from this logic synthesis step is then fed into a physical implementation process, which includes the place and route steps.

Logic synthesis is the process of transforming the chip's RTL description into a technology-dependent gate netlist by using the library's logical view. In contrast to RTL description, which only contains functional information, the gate netlist is the standard cell representation of the design at the component level. It is comprised of gate instances and the port connectivity among these instances. The primary requirement for the task of logic synthesis is ensuring the mathematical equivalency between the synthesized gate netlist and the original RTL description.

The process of *place* is the first step in creating the chip in a physical domain. It determines the physical locations of each individual cell in the netlist based on design constraints. Placement is a complicated process that is very algorithm intensive and time-consuming. The quality of the placement work has a preeminent impact on the chip's performance. The following *route* process is also critical. It creates the physical wire connections for the signal and power nets that are defined in the logic connectivity of the netlist. It is a very complicated process whose goals include meeting the design speed target, minimizing the total wire length, and avoiding the design rule violations.

After the place and route steps, the resultant physical entity is checked against various rules, such as the process manufacturing rules (foundry design rules) and design integrity and reliability criteria. This physical entity is also checked logically to ensure that it matches the design intention defined in the original RTL code. After these rigorous checks, the final layout is sent to the mask shop for the creation of photomasks. This is called *tapeout,* the final step in this standard cell-based ASIC (application-specific integrated circuit) design approach.

Currently, the standard cell-based ASIC approach is the main design methodology for commercial products, especially for large digitally dominated designs. A majority of SoC projects are carried out with this implementation approach.

8. WHAT IS THE SYSTEM-ON-CHIP (SOC) APPROACH?

SoC or *system on chip* is the design approach of integrating the components of an electronic system into a single chip. In the past, chips could only perform dedicated simple functions, such as simple logic operations, decoding/encoding operations, analog-to-digital conversion, digital-to-analog conversion, and so on. As time went by, more and more functions were integrated into a single chip. This integration trend is so significant that it has reached the point where a single chip can perform the functions of an entire electronic system, such as an MPEG decoder, a network router, or a cellular phone. As a result, a colorful name was created for such chips: system on chip (SoC). SoC designs often consume less power, cost less, and are more reliable than the multichip systems that they are designed to replace. Furthermore, assembly cost is reduced due to the fact that there are fewer packages in the system.

The key to the SoC approach is integration. By integrating increasingly more preassembled and verified blocks, which have dedicated functions, into one chip, a sophisticated system is created in a timely and economical fashion. Figure 1.3 is a block diagram of a SoC that shows the various blocks on a chip. As seen in the figure, integrating predesigned and verified blocks into a large chip is the essence of SoC approach.

A typical SoC chip has one or more microprocessors or microcontrollers on board, the brain of the SoC chip. The on-chip processor (e.g., an RISC controller) coordinates the activities inside the chip. In some cases, a dedicated DSP engine, which targets algorithm-intensive signal processing tasks, may also be found on a SoC chip. Having a large number of memory blocks is another characteristic of a SoC chip. These mem-

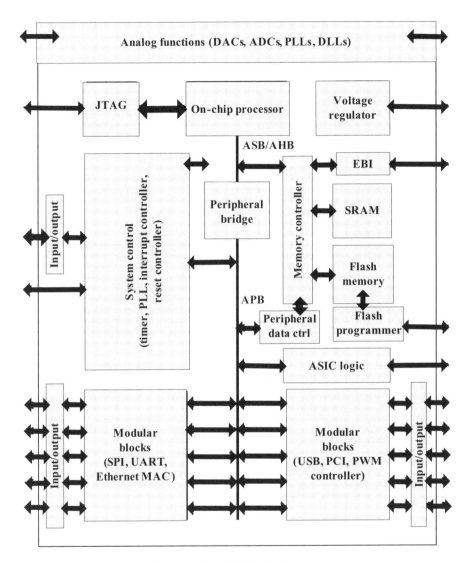

Figure 1.3. A SoC block diagram.

ories (ROM, RAM, EEPROM, and Flash) support the SoC's software functions. Another indispensable component of a SoC chip is the timing source, which includes an oscillator and phase lock loop (PLL). It is almost always true that one or more PLLs are found on any SoC chip since most SoC designs are based on synchronous design principle, and clocks are the key design feature.

A SoC needs external interfaces, such as industry standard USB, Firewire, Ethernet, and UART, to communicate with the outside world. A direct memory access (DMA) controller can be used to route data directly between the external interfaces and memories, bypassing the on-chip processor and thereby increasing the data throughput.

If a SoC is designed to interface with devices that have direct contact with human activities, some analog components, such as ADC or DAC, are essential. In some cases, on-chip voltage regulators and power management circuits can be found in a SoC as well.

To tie the components of a SoC together, an on-chip bus architecture is required for internal data transferring. This is either a proprietary bus or an industry-standard bus such as the AMBA bus from ARM. *Network on a chip (NoC)* is a new approach to SoC design. In an NoC system, modules such as processor cores, memories, and specialized IP blocks exchange data using a network as a public-transportation subsystem. The network is constructed from multiple point-to-point data links interconnected by switches such that messages are relayed from any source module to any destination module over several links by making routing decisions at the switches.

The NoC approach brings a networking solution to on-chip communication and provides notable improvements over conventional bus systems. From the viewpoint of physical design, the on-chip interconnect dominates both the dynamic power dissipation and performance of deep submicron CMOS technologies. If a signal is required across the chip, it may require multiple clock cycles when propagated in wires. A NoC link, on the other hand, can reduce the complexity of designing long interconnecting wires to achieve predictable speed, low noise, and high reliability due to its regular, well-controlled structure. From the viewpoint of system design and with the advent of multicore processor systems, a network is a natural architectural choice. A NoC can provide separation between the tasks of computation and communication, support modularity, and IP reuse via standard interfaces, efficiently handle synchronization issues, and serve as a platform for system test.

Just as the major hardware blocks are critical, so is the software of a SoC. The software controls the microcontroller, microprocessor, and DSP cores; the peripherals, and the interfaces to achieve various system functions.

One indispensable step in SoC development is *emulation*. Emulation is the process of using one system to perform the tasks in exactly the same way as another system, perhaps at a slower speed. Before a SoC device is sent out to fabrication, it must be verified by emulation for behavior analysis and making predications. During emulation, the SoC hardware is mapped onto an emulation platform based on a FPGA (or the likes) that mimics the behavior of the SoC. The software modules are loaded into the memory of the emulation platform. Once programmed, the emulation platform enables both the testing and the debugging of the SoC hardware and the software.

In summary, the SoC approach is primarily focused on the integration of predesigned, preverified blocks, not on the design of individual components. In other words, the keyword is *integration,* not design.

9. WHAT ARE THE DRIVING FORCES BEHIND THE SOC TREND?

One of the major driving forces behind the SoC trend is cost. Integrating more functions into a single chip can reduce the chip count of a system and thus shrink the package and board cost. It could potentially lower the overall system cost and make the product more competitive. In today's consumer electronic market and in others, better price always provides advantage of gaining market share. During the past decade (from the late 1990s) or so, the SoC approach has been proven to be one of the most effective ways of reducing the cost of electronic devices.

The other forces behind this trend include pursuing higher chip performance or higher operating frequency. This is owing to the fact that SoC can eliminate interchip communication and shorten the distances among the on-chip components, which positively enhances the chip speed. In some cases, the demand for overall lower system power usage is also a factor for choosing the SoC approach. And, portability is another advantage of the SoC method. When a system is migrated from an old process to a new one, SoC can greatly reduce the workload compared to the transfer of several chips.

Overall, SoC chip implementation has enabled many technology innovations to reach the consumer in shorter and shorter time frames.

10. WHAT ARE THE MAJOR TASKS IN DEVELOPING A SOC CHIP FROM CONCEPT TO SILICON?

The process of developing a SoC chip from concept to silicon is divided into the following four tasks: design, verification, implementation, and software development.

Design often starts with marketing research and product definition and is followed by system design. It ends with RTL coding.

Verification is a means of ensuring that the chip can perform faithfully in functionality, according to its design specifications. It includes verification at the system, RTL, and gate levels, and sometimes even at the transistor level. This bug-finding struggle continues until the chip is ready to ramp into production.

Implementation is the process of actually creating the hardware, which results in an entity that one can see and feel. It includes both the logical and physical implementations.

Software development is the process of programming the brain of the SoC (the on-chip processors), or arming the chip with intelligence.

11. WHAT ARE THE MAJOR COSTS OF DEVELOPING A CHIP?

There are two types of costs associated with the task of developing a VLSI chip: fixed costs and variable costs.

Fixed costs are also called *nonrecurring engineering (NRE)* costs. These refer to the one-time costs of researching, designing, and testing a new product. When developing a budget for a project and analyzing if a new product will be profitable, NRE must be considered. In the chip designing business, these costs include the engineering design cost (salaries, EDA tools licenses, CPU time, disk space, etc.) and mask (reticle) cost. Currently (2006), for 90 nm technology, the mask cost alone is in the neighborhood of one million dollars. This fixed cost is nonrefundable and is unrelated to product volume. To achieve profitability, then, the product must sell well enough to produce a return that covers at least the initial NRE and the materials and processing costs to make the initial material.

Variable cost is the cost of manufacturing, testing, and packaging the production chip. This cost is proportional to volume as every chip needs raw material for manufacture and tester's time for testing. This expense must be paid continually in order to maintain the product's manufacture.

The NRE or fixed cost represents a significant percentage of the overall cost of small-volume products. As volume grows, however, the fixed cost is gradually buried below the surface and the variable cost becomes dominant.

During the early phase of a project, the cost study is a very important, complex, and sensitive subject. Predicting or estimating the downstream expense and potential revenue-generating capability of a product with reasonable accuracy is not a trivial task. Sometimes, it is very hard to make a decision whether to kill or support a technically promising but market-unfriendly project as there might always be unknowns. The market is dynamic in nature; something unattractive today could be very popular tomorrow and vice versa.

CHAPTER *2*

The Basics of the CMOS
Process and Devices

12. WHAT ARE THE MAJOR PROCESS STEPS IN BUILDING MOSFET TRANSISTORS?

VLSI chips are manufactured on semiconductor material whose electrical conductivity lies between that of an insulator and a conductor. The electrical properties of semiconductors can be modified by introducing impurities through a process known as doping. The ability to control conductivity in small and well-defined regions of semiconductor material has led to the development of semiconductor devices. Combined with simpler passive components (resistors, capacitors, and inductors), they are used to create a variety of electronic devices.

As addressed in Chapter 1, Question 4, CMOS is the most popular semiconductor process. The *metal oxide semiconductor field effect transistor (MOSFET)* is the transistor of the CMOS process. MOSFETs are comprised of channels of n-type or p-type semiconductor material and are thus called NMOSFET and PMOSFET, or NMOS and PMOS for short. The MOSFET has emerged as the omnipresent active element for the VLSI integrated circuit. The competitive drive for better performance and reduced cost has resulted in the scaling of circuit elements to smaller and smaller dimensions.

MOSFET transistors are built through the semiconductor fabrication process, which is a sequence of multiple photographic and chemical processing steps. The electronic circuits are gradually, in this step-by-step manner, created on a wafer of pure semiconductor material such as silicon.

In semiconductor fabrication, the various processing steps are grouped into four general categories: deposition, removal, patterning, and modification of electrical properties. Deposition is any process that grows, coats, or transfers a material onto the wafer. Available deposition technologies are *physical vapor deposition (PVD), chemical vapor deposition (CVD), electrochemical deposition (ECD), molecular beam epitaxy (MBE),* and *atomic layer deposition (ALD).*

Removal processes are techniques for removing material from the wafer either in bulk or selectively. The primary removal method is etching, including both wet etching and dry etching. *Chemical–mechanical planarization (CMP)* is another removal technique that is used between different processing levels. Patterning is the series of processes that pattern or alter the existing shape of the deposited materials and is generally referred to as *lithography*. Modifying electrical properties consists of doping a transistor's source and drain in diffusion furnaces or by implanting it with ions. These doping processes are followed by furnace annealing or rapid thermal annealing (RTA), which activates the implanted dopants. The modification of electrical properties now also includes the reduction of the dielectric constant in low-k insulating materials via exposure to ultraviolet light.

Modern processes often have more than twenty mask levels with more than one hundred processing steps. These steps are classified as either *front-end* or *back-end* processing. Front-end processing refers to the formation of the transistors directly on the silicon. Back-end processing is the creation of metal interconnecting wires, which are isolated by insulating materials, to connect the transistor formations. The processing steps are grouped roughly as follows:

- Silicon crystal growth
- Wet cleaning
- Photolithography
- Ion implantation
- Dry etching
- Wet etching
- Plasma etching
- Thermal treatments (rapid thermal annealing, furnace annealing, and oxidation)
- Chemical vapor deposition
- Physical vapor deposition
- Molecular beam epitaxy
- Electrochemical deposition
- Metallization
- Chemical–mechanical planarization
- Wafer testing
- Wafer backgrinding
- Wafer mounting
- Die cutting

13. WHAT ARE THE TWO TYPES OF MOSFET TRANSISTORS?

There are two types of MOSFET transistors: NMOS and PMOS. The transistors are called *active* devices because they can convert and amplify voltage and current. The inductors, resistors, and capacitors are called passive devices because they only consume energy. All of the functions of a VLSI CMOS chip, no matter how complex it is, are achieved by these two types of transistors with the aid of the passive devices. Figure 2.1 shows the symbols of PMOS and NMOS transistors. G, D, and S signify the gate, drain and source terminals of the transistor.

In an n-well process, the circuits are built on a p-type wafer. In this type of process, the NMOS transistors are directly fabricated on the wafer, whereas the PMOS transistors are fabricated in the n-well. Figure 2.2 is the cross section of a wafer where NMOS and PMOS transistors are fabricated. The p-substrate is often tied to the lowest voltage of the chip; the n-well is tied to the highest voltage. The substrate and well are often referred to as the body or bulk of the MOSFET. Both n and p MOSFETs are four-terminal devices with terminals of source, drain, gate, and body or bulk. When body and bulk are tied to V_{DD} and V_{SS}, the fourth terminals (body and bulk) are not drawn, as in the case of Figure 2.1.

The gate terminal is a layer of polysilicon placed over the channel. It is separated from the channel by a thin insulating layer of silicon dioxide or silicon oxynitride. When a voltage is applied between the gate and source terminals, the electric field that is generated penetrates the oxide and creates an *inversion channel* in the channel underneath. The inversion channel is of the same type (n or p) as the source and drain and provides a conduit through

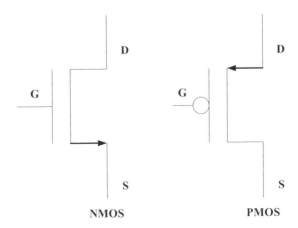

Figure 2.1. Symbols for NMOS and PMOS transistors.

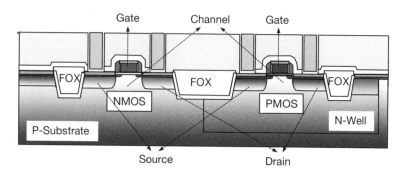

Figure 2.2. A cross section of NMOS and PMOS transistors.

which electric current can flow. Varying the voltage between the gate and source modulates the conductivity of the channel and, consequently, makes it possible to control the current flow between the drain and source terminals.

The operation of a MOSFET transistor is classified into three different modes, depending on the voltages at the gate, source, and drain terminals. The cutoff (or subthreshold) mode refers to the turned-off MOSFET, where there is no conduction between drain and source. The linear (or triode) mode refers to the creation of a channel that allows current to flow between the drain and source. In this mode the transistor is turned on, and the MOS-FET operates like a resistor since the source–drain current is linearly proposional to the drain voltage (at a given gate voltage). The saturation mode refers to operation in which the drain voltage is higher than the gate voltage and, thus, a portion of the channel is turned off. When a transistor enters the saturation mode, the source–drain current is relatively independent of the drain voltage and is controlled only by the gate voltage. These three operating modes (regions) are depicted in Figure 2.3. Using these modes, digital as well as analog circuits are constructed.

The physics of MOSFET transistors have been studied intensively by process scientists and engineers. From this research, models have been created to describe MOSFET behavior under different conditions. Circuit designers can then use these models in their circuit simulations. The most popular model is the *Berkeley short-channel Igfet model (BSIM),* which is used in SPICE simulators for describing the n- and p-channel MOS transistor behavior.

14. WHAT ARE THE BASE LAYERS AND METAL LAYERS?

As discussed in Question 13, base layers are used in front-end processes to create NMOS and PMOS transistors, whereas metal layers are used in back-

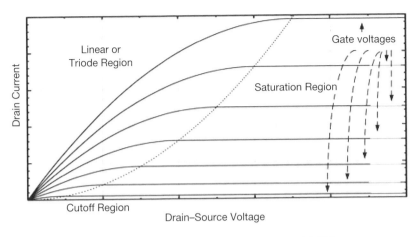

Figure 2.3. The three operating modes of a MOSFET.

end processes to connect these transistors. After tapeout, each of these layers requires a dedicated photomask for its patterns.

The base layers are comprised primarily of the following:

- A n-well to define the n-well area, an implanted or diffused region in the silicon wafer.
- An active area to define the region for n- and p-channel devices.
- Poly to define the gates of the devices.
- Cont. This is an open area in active area and poly for connecting to metal1.
- A tap. This identifies the n-well and substrate cont.

In semiconductor process technology, metal refers to a material with very high electrical conductivity. When voltage is applied at the ends of metal, electrons can move around almost freely within the metal. Due to its high conductivity, metal connects on-chip devices. Aluminum is the most commonly used metal; it is the most process-friendly metal and has low resistivity. Its shortcomings include electromigration and insufficient temperature resistance. Copper is the metal of choice for advanced processes. Its resistivity is the lowest among the metals. Compared to aluminum, it has lower resistivity and lower electromigration. However, it is harder to process.

Interconnects in high-density IC chips are formed by multilevel networks, as seen in Figure 2.4. For a typical 90 nm CMOS process, there could be seven or eight levels of metals. Between any two adjacent metal levels, there is a dedicated layer called *via* that is used to make the necessary

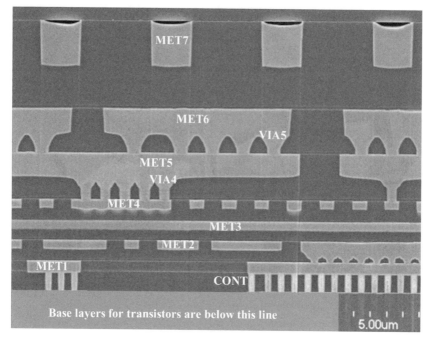

Figure 2.4. SEM cross section showing metal levels.

connection between the two metals. A via is a hole etched in the interlayer dielectric, which is then filled with metal, usually tungsten, to provide a vertical connection between stacked interconnect metal lines. The group of metal layers is comprised of the following:

- Metal1, the first level of interconnect
- Via1, to connect the metal1 and metal2
- Metal2, the second level of interconnect
- Via2, to connect the metal2 and metal3
- Metal3, the third level of interconnect
- Via3, to connect the metal3 and metal4
- Metal4, the fourth level of interconnect
- Via4, to connect the metal4 and metal5
- Metal5, the fifth level of interconnect
- Via5, to connect the metal5 and metal6
- Metal6, the sixth level of interconnect
- Via6, to connect the metal6 and metal7
- Metal7, the seventh level of interconnect

Figure 2.5 shows the detail of a metal cross section. W is the metal width, H is metal height, and d is the distance between two adjacent, same-layer metal pieces. For a metal piece of length L, the resistance associated with it can be expressed as $R = \rho \cdot (L/A)$, where ρ is the metal conductivity and A is the area of the cross section, or $A = W \cdot H$. The minimum allowable metal width W and the minimum distance d (metal pitch) are critical process metrics. They are scaled accordingly as CMOS transistors scale down from generation to generation. The gate density of the process is closely related to these parameters. As shown in the figure, in a newer process, the metal width W is reduced. Consequently, the associated R increases. To diminish this effect, the metal height H is increased; this results in the high-aspect-ratio metal cross section. Along with reduced distance d, these high-aspect-ratio metals form perfect parallel-plate capacitors, which aggravate cross talk between the electrical signals traveling through these metal lines.

As the semiconductor industry advances to even finer process geometries, the problems associated with interconnection will become more difficult. Most challenging for interconnection is the introduction of new materials that meet the requirements of wire conductivity and reduced dielectric permittivity (a measure of the ability of a material to resist the formation of an electric field within it). New materials, structures, and processes also create new reliability problems (electrical, thermal, and mechanical). Detecting, testing, modeling, and controlling failure mechanisms will be the key to solving these problems.

Dimensional control is another challenge for present and future interconnect technologies. To extract maximum performance, interconnect struc-

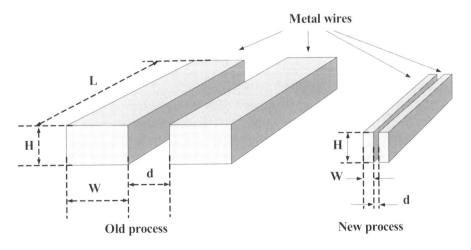

Figure 2.5. Metal cross section detail.

tures cannot tolerate profile variability without producing undesirable RC degradation. This requires tight control of pattern, etch, and planarization. The control of the interconnect features must be three-dimensional to achieve the necessary circuit performance and reliability. These dimensional control requirements place new demands on high-throughput imaging metrology for measuring high-aspect-ratio structures. New metrology techniques are also needed for in-line monitoring of adhesion and defects.

As feature sizes shrink, interconnect processes must be compatible with device road maps and meet the manufacturing targets of the specified wafer sizes. In summary, the challenges of current and future interconnect technologies—plasma damage, contamination, thermal budgets, cleaning of high-aspect-ratio features, and defect tolerant processes—are key practical concerns in determining manufacturability and in the control of defects while meeting overall cost and performance requirements.

15. WHAT ARE WAFERS AND DIES?

Silicon is the most essential semiconductor material used in solid state electronics. Silicon in the form of single-crystal wafer is the basic building block for IC fabrication. To keep pace with the growth in IC processing technology, chip size, and circuit complexity, a silicon crystal and a wafer must be prepared with increasing diameters and improved quality.

A wafer is the circular silicon base upon which chips are manufactured. It is made from an ingot, which is a cylindrical, single-crystal semiconductor typically resulting from the Czochralski crystal growth process, as depicted in Figure 2.6. During this process, high-purity silicon is melted down in a crucible (made of quartz). Dopant impurity atoms such as boron or phosphorus may be added to the molten intrinsic silicon in precise amounts to dope the silicon, thus changing it into n-type or p-type extrinsic silicon. This may influence the electrical conductivity of the silicon. A seed crystal mounted on a rod is dipped into the molten silicon. This seed crystal rod is continuously pulled upwards and rotated at the same time. The crystal ingot is then built layer by layer of atoms. By precisely controlling the temperature gradients, the rate of pulling, and the speed of rotation, a large single-crystal, cylindrical ingot can be extracted from the melt. This process is normally performed in an inert atmosphere (such as argon) and in an chamber made of an inert matherial (such as quartz). Figure 2.7 is a photo of a silicon ingot.

Using high-precision diamond saws or diamond wires, the ingot is first shaped and then sliced into wafers with thicknesses on the order of 0.5 mm,

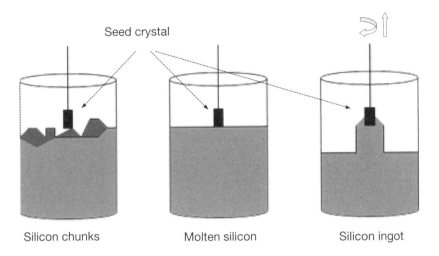

Seed crystal

Silicon chunks   Molten silicon   Silicon ingot

Figure 2.6. The crystal growth process.

as shown in Figure 2.8. This wafer fabrication process includes the steps of cutting, grinding, polishing, and cleaning to transform a single-crystal rod into many circular wafers for manufacture into semiconductor devices. A wafer is measured by its diameter: 4 inches, 6 inches, 8 inches, or 12 inches. There is no plan for wafers larger than 12 inches in the near future due to the technological difficulties of handling super-large-sized silicon "pizzas."

Figure 2.7. A silicon ingot.

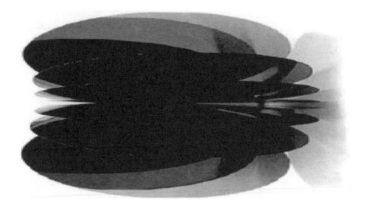

Figure 2.8. Wafers cut from a silicon ingot.

Inside a wafer, as shown in Figure 2.9, there are many small blocks or cells. These individual cells are called *dies* or *chips*. A die is a small piece of silicon material upon which a given circuit is fabricated. Typically, integrated circuits are produced in large batches on a single wafer through various process steps (see also Questions 12 and 13). The resultant wafer is then cut into pieces, each containing one copy of the desired integrated circuit. Each one of these pieces is a die.

Die cutting, or dicing, is the process of separating a wafer of multiple identical integrated circuits into dies, each containing one of those circuits. As seen in Figures 2.9 and 2.10, there is a thin, nonfunctional space between

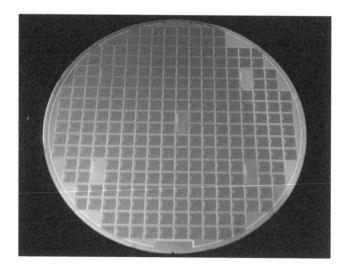

Figure 2.9. A wafer with dies.

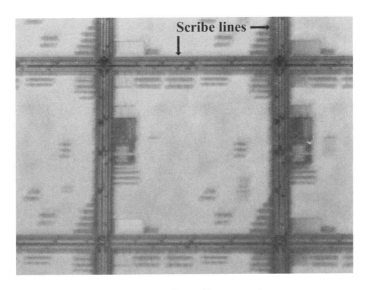

Figure 2.10. A die or chip on a wafer.

the functional parts of the circuits in which a saw can safely cut the wafer without damaging the circuit. This spacing is called the scribe line. The width of the scribe line is very small. A very thin and accurate saw is therefore required to cut the wafer into pieces. The dicing is performed with a water-cooled circular saw that has diamond-tipped teeth, as graphically demonstrated in Figure 2.11.

As explained, the dicing process is usually accomplished by mechanical sawing. However, laser cutting is another technique. In laser cutting, a high-powered laser beam is directed at the material. As a result, the material either melts, burns, or vaporizes away. The advantages of laser cutting over mechanical cutting include the lack of physical contact (no material contam-

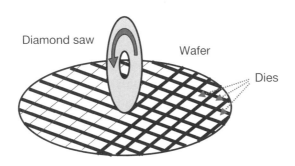

Figure 2.11. Diamond sawing.

ination), a high level of precision (there is no wear on the laser), and the reduced risk of warping the material that is being cut as it produces a reduced heat-affected zone. The disadvantage of laser cutting is primarily the high energy required.

Following the dicing process, the individual silicon chips are encapsulated into packages, which are then suitable for building electronic devices.

16. WHAT IS SEMICONDUCTOR LITHOGRAPHY?

As discussed in Questions 12, 13, and 14, the fabrication of an integrated circuit requires a variety of physical and chemical processes to be performed on a silicon substrate. In general, the various processes fall into four categories: film deposition, removal, patterning, and doping. Films of both conductors (polysilicon, aluminum, and copper) and insulators (various forms of silicon dioxide, silicon nitride, and others) connect and isolate transistors of the circuit. Selective doping of various regions in the silicon changes the conductivity of the affected regions. By performing the various process steps in certain sequences, millions of transistors can be built and wired together to form the complex circuitry of a modern electronic device. Fundamental to these processes is *lithography,* which is the formation of three-dimensional images on the substrate for subsequent transferring of patterns to the substrate.

Semiconductor lithography is a process of drawing patterns on a silicon wafer. The patterns are drawn with a light-sensitive polymer called photoresist. To build the complex structures required for an integrated circuit, the lithography and etch-pattern transferring steps are typically repeated 20–30 times. Each pattern printed on the wafer must align with the previously formed patterns. Step-by-step, the conductors, insulators, and selectively doped regions are slowly built up to form the final devices.

The overwhelming technology choice for performing this patterning is optical lithography. It is basically a photographic process by which the light-sensitive polymer (photoresist) is exposed and developed to form three-dimensional images on the substrate. Ideally, the photoresist image should have the exact shape of the intended pattern in the plane of the substrate. The final photoresist pattern is binary: parts of the substrate are covered with resist while other parts are completely uncovered. This binary pattern is necessary for pattern transfer as the parts of the substrate covered with resist will be protected from etching, ion implantation, or other pattern transfer mechanisms.

The photomask is an essential component in semiconductor lithography. It contains the detailed blueprint of the designed circuit. Using the photomask, specific images of detailed devices are transferred onto the surface of the silicon wafers by means of photolithography. The principle of photomasking is similar to photography in many ways. A photomask is used just like the negative in photography that captures specific images for later reproduction. In photography, multiple copies of photos are reproduced using the original image captured on the negative. Likewise, a photomask produces duplicate images or patterns onto the silicon wafers. A single photomask plate produces identical images on thousands of wafers. As the quality of the finished photograph is determined by the quality of the original film, the quality of the photomask determines the ultimate quality of semiconductor chips.

The material used for building photomasks is a quartz plate upon which detailed images or patterns are formed. The patterns or images are then transferred onto the wafer surfaces by shining light through the quartz plate as depicted in Figure 2.12, just as negative film projects images onto photographic paper.

In the mass production of integrated circuits, photomasks are also called photoreticles, or reticles for short. Reticles are created by very complicated and expensive machines. Each reticle contains only one layer of the circuit. A set of reticles, each defining one pattern layer, is fed into a photolithography stepper or scanner and individually selected for exposure to form the desired pattern on the wafer. Circuit elements (transistors, capacitors, and

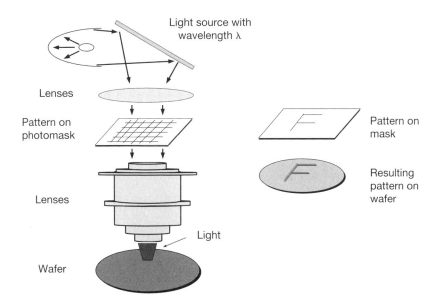

Light source with wavelength λ

Lenses

Pattern on photomask

Lenses

Light

Wafer

Pattern on mask

Resulting pattern on wafer

Figure 2.12. The optical lithography system.

resistors) are created by those patterns of many layers. A complete design can require more than 20 masks in a modern CMOS process. Figure 2.13 is the metal1 mask of a 0.35 μm technology.

The general step sequence for a typical optical lithography process is substrate preparation, photoresist spin coating, prebaking, exposure, post-exposure baking, development, postbaking, and etching and implanting. After the final step of photoresist stripping, the pattern is transferred to the underlying layer. This sequence of steps is shown graphically in Figure 2.14.

To expose the photoresist through the photomask, several techniques have been used during the history of the lithography: contact printing, proximity printing, and projection printing. Contact lithography offers high resolution. But practical problems such as mask damage and the resultant low yield make this process unusable in most production environments. Proximity printing reduces mask damage by keeping the mask a small distance above the wafer. However, the resolution limit is increased, making proximity printing insufficient for today's technology.

By far the most common method of exposure is projection printing. Projection lithography derives its name from the fact that an image of the mask is projected onto the wafer, as shown in Figure 2.12. Projection lithography gradually became a valuable alternative to contact and proximity printing in the mid 1970s when the advent of computer-aided lens design and improved optical materials allowed the production of lens elements of sufficient quality to meet the requirements of the semiconductor industry. There are two major classes of projection lithography tools: scanning and step-and-repeat

Figure 2.13. A metal1 mask.

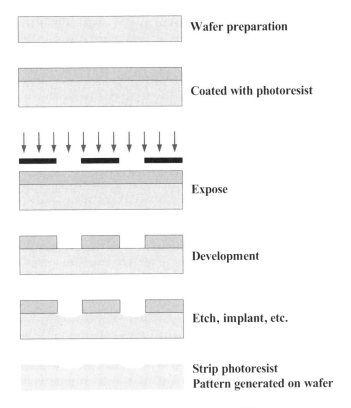

Wafer preparation

Coated with photoresist

Expose

Development

Etch, implant, etc.

Strip photoresist
Pattern generated on wafer

Figure 2.14. The sequence of major steps in a typical lithography process.

systems. Scanning projection printing employs reflective optics to project a slit of light from the mask onto the wafer as the mask and wafer are moved simultaneously past the slit. Exposure dose is determined by the intensity of the light, the slit width, and the speed at which the wafer is scanned. These early scanning systems are 1:1 (the mask and image sizes are equal). The step-and-repeat system, or stepper for short, exposes the wafer one rectangular section (image field) at a time and can be 1:1 or a reduction. Stepper systems employ refractive optics (lenses) and are quasimonochromatic. Both scanning and step-and-repeat systems are capable of high-resolution imaging. Reduction imaging is required for the highest resolution. Figure 2.15 diagrammatically shows the idea of the scanner and step-and-repeat system.

The step-and-repeat system continued to dominate semiconductor lithographic patterning throughout the 1990s as minimum feature sizes reached the 250 nm levels. However, by the early 1990s a hybrid step-and-scan approach was introduced. The step-and-scan approach uses a fraction of the

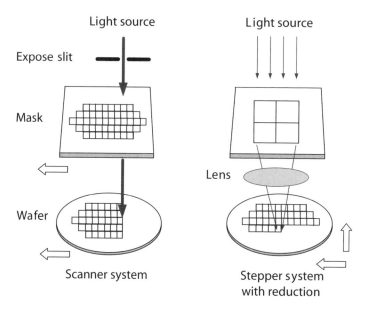

Figure 2.15. The scanner and step-and-repeat system.

normal stepper field and then scans this field in one direction to expose the entire 4× reduction mask. The wafer is then stepped to a new location, and the scan is repeated. The smaller imaging field simplifies the design and manufacturing of the lens, but at the expense of a more complicated reticle and wafer stage. Step-and-scan technology is the technology of choice today for < 250 nm manufacturing.

In semiconductor lithography, resolution is the smallest feature that can be printed on a wafer with adequate control. It has two basic limits: the smallest image that can be projected onto the wafer and the resolving capability of the photoresist to make use of that image. From the imaging projection side, resolution is determined by the wavelength of the imaging light. During various periods, photomasks have used wavelengths of 365 nm and 248 nm; today's mainstream high-resolution wavelength is 193 nm. High-index immersion lithography is the newest extension of 193 nm lithography being considered.

Other alternatives are *extreme ultraviolet lithography (EUV), nanoimprint lithography,* and *contact printing.* EUV lithography systems that will use 13.5 nm wavelengths, approaching the regime of X-rays (10–0.01 nm), are currently under development. Nanoimprint lithography is being investigated as a low-cost, nonoptical alternative. Contact printing may be revived with the recent interest in nanoimprint lithography.

After the small patterns have been lithographically printed in photoresist, they must be transferred onto the substrate. There are three basic pattern transferring approaches: subtractive transfer (etching), additive transfer (selective deposition), and impurity doping (ion implantation). Etching is the most common approach. A uniform layer of the material for patterning is deposited on the substrate. Lithography is then performed such that the areas for etching are left unprotected (or uncovered) by the photoresist. Etching is performed using either wet chemicals, such as acids, or in a dry plasma environment. The photoresist resists the etching and protects the material covered by the photoresist. When the etching is complete, the photoresist is stripped, leaving the desired pattern etched into the deposited layer (see Figure 2.14).

Additive processes are used whenever workable etching processes are not available. In this method, the lithographic pattern is applied where the new layer is to be grown (by electroplating in the case of copper). Stripping of the photoresist then leaves the new material as a negative version of the patterned photoresist. Doping involves the addition of controlled amounts of contaminants that change the conductive properties of a semiconductor. Ion implantation uses a beam of dopant ions directed at the photoresist-patterned substrate. The photoresist blocks the ions, but the areas uncovered by resists are embedded with ions. This can create the selectively doped regions that make up the electrical heart of the transistors.

In summary, the importance of semiconductor lithography can be appreciated in two ways. First, due to the large number of lithography steps needed in IC manufacturing, lithography typically accounts for large percentage of the chip's manufacturing cost. Second, lithography tends to be the technical limiter for further advances in feature size reduction (and consequently, the transistor speed and silicon area). Although lithography is not the only technically critical and challenging process in the IC manufacturing flow, historically, advances in lithography have guided the advances in integrated circuit's performance and cost.

17. WHAT IS A PACKAGE?

A *package* is the housing of a semiconductor chip. It protects and preserves the performance of the semiconductor devices from electrical, mechanical, and chemical corruption or impairment. It also electrically interconnects the chip with outside circuitry. Further, it is designed to dissipate heat generated by the chip. Package design is becoming increasingly significant as well as difficult as device performance, complexity, and functionality increase with each generation of technology.

Structurally, a package is a plastic, ceramic, laminate, or metal seal that encloses the chip or die inside, as depicted in Figure 2.16. The package can protect the chip from contamination or damage by a foreign material in the environment. It is a medium to host the chip or die on the printed circuit board (PCB). Packages can be classified into two categories, according to the way in which they attach to the PCB. In *pin-through-hole (PTH)* packages, pins are inserted into through-holes in the board and soldered in place from the opposite side of the board. *Surface-mount-technology (SMT)* packages have leads that are soldered directly to the metal leads on the surface of the circuit board. SMT packages are generally preferred.

During the IC packaging process, the following operations are performed: die attaching, bonding, and encapsulation. Die attaching is the step during which a die is mounted and fixed to the package or support structure. Bonding is the process of creating interconnections between the die and the outside world. Encapsulation refers to the die being encapsulated with ceramic, plastic, metal, or epoxy to prevent physical damage or corrosion. This term is sometime used synonymously with "packaging."

There are two different approaches for IC bonding: wire bonding and flip-chip bonding. Figure 2.17 shows a photo of a wire-bonded package. Wire bonding is the technique of using thin wire and a combination of heat, pressure, and/or ultrasonic energy to make the interconnection between the die and package. The wires are made of gold, aluminum, or copper. Wire diameters start at 15 μm and may increase to several hundred micrometers for high-powered applications. There are two main classes of wire bonding: ball bonding and wedge bonding. In both types, the wire is attached at the ends using some combination of heat, pressure, and ultrasonic energy to make a weld. Figure 2.18 shows the contact points for both ball and wedge bonding. Wire bonding is generally considered the most cost-effective technology, and it is currently used to assemble the vast majority of semiconductor packages.

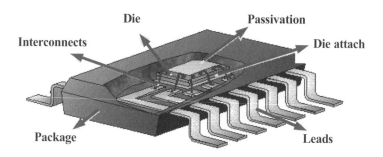

Figure 2.16. A die inside a package.

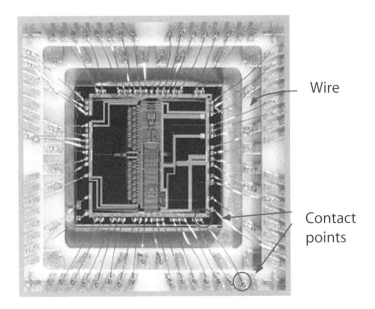

Figure 2.17. A wire-bonded package.

Ball bonding   **Wedge bonding**

Figure 2.18. Ball and wedge bonding.

Flip-chip bonding is the type of mounting technique that does not require any wire bonds. It is the direct electrical connection between the face-down die and the substrate or circuit board, through the conductive bumps on the chip's bond pads. This idea is depicted in Figure 2.19. The space between the die and the substrate is filled with a nonconductive *underfill* adhesive. The underfill protects the bumps from moisture or other environmental hazards and provides additional mechanical strength to the assembly. However, its most important purpose is to compensate for any thermal expansion difference between the chip and the substrate. The underfill mechanically locks together the chip and the substrate so that differences in thermal expansion do not break or damage the electrical connection of the bumps.

The advantages of the flip-chip technique include small size, high performance, great flexibility, and high reliability. Flip-chip packages are smaller because the package and bond wires are eliminated, thus reducing the required board area and resulting in far less height. Weight is reduced as well. By eliminating the wire bonds, the signal paths are significantly shortened and, thus, greatly trim the delaying inductance and capacitance of the interconnection. This results in high-speed off-chip interconnections and, thus, high performance. In the case of wire bonds, the I/O connections are limited to the perimeter of the die, driving the die sizes up as the number of connections increases. Flip-chip connections can use the whole area of the die, accommodating many more connections on a smaller die. Mechanically, the flip-chip connection is the most rugged interconnection method.

The major "care-abouts" for package design include thermal, electrical, mechanical, and cost. The power consumed by a chip is converted into heat. This generated heat causes a rise in temperature. A semiconductor device operates normally as long as the temperature does not exceed an upper limit, which is specified as the ambient temperature and the temperature of the junctions inside the device. When this upper limit is exceeded, the semicon-

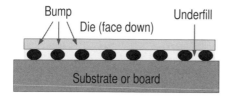

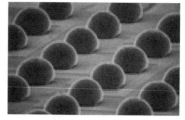

Bumps on die

Figure 2.19. Flip-chip bonding.

ductor device stops operating normally and becomes damaged. Therefore, it is necessary to successfully dissipate the generated heat so as to keep the temperature within the specified limits.

The thermal management of the package is crucial since it must transfer the generated heat to outside world quickly and efficiently. As a chip's gate density increases with CMOS scaling, this problem becomes more challenging since high-density design consumes more power, and smaller systems have reduced airflow and heat-sinking capabilities. A parameter of thermal resistance is defined as $\Theta_{xy} = (T_x - T_y)/P$, where T_x and T_y are the temperatures of any two points in the package environment and P is the power usage of the chip.

There are three basic temperatures in package design: junction temperature T_j, case temperature T_c, and ambient temperature T_a. Correspondingly, the three crucial thermal resistances are Θ_{jc}, Θ_{ca}, and Θ_{ja}. The thermal resistance between the junction and the case is determined by the structure of the device and the material of the package, and it is a fixed value after the chip and package are determined. However, the thermal resistance between the case and the ambient air can vary greatly according to the mounting conditions. For example, radiator attachment is often used to route the heat from the case surface to ambient air for high-power designs. Typically, maximum junction temperature is one of the requirements in the specification of a package design. Currently, 105°C is the recommended maximum junction temperature for advanced CMOS technology.

Electrically, the need for greater and faster interconnects to and from the chip and the need for more effective management of the mixed signals and high-frequency behavior are critical factors in package development. In addition to these considerations is packaging cost. Packaging cost is a significant part of overall chip cost. It must be treated seriously during the planning and cost analysis stages of a chip project. It should be as low as possible for a better profit margin.

The commonly used package types appear in Figures 2.20 through 2.29.

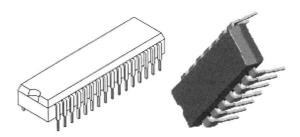

Figure 2.20. DIP, dual in-line package.

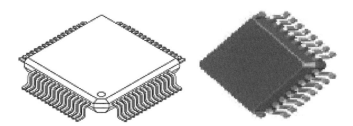

Figure 2.21. QFP, quad flat package.

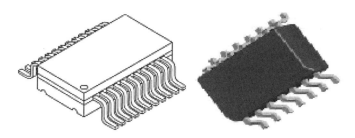

Figure 2.22. SQIC, small outline IC.

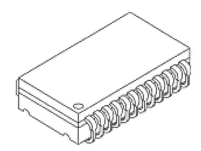

Figure 2.23. SOJ, small outline J-leaded.

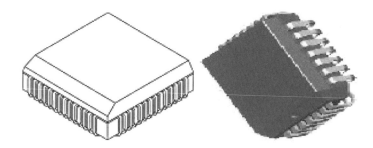

Figure 2.24. PLCC, plastic leadless chip carrier.

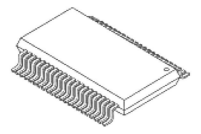

Figure 2.25. TSOP, thin small outline package.

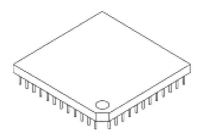

Figure 2.26. PGA, pin grid array.

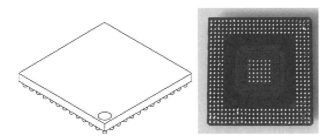

Figure 2.27. BGA, ball grid array.

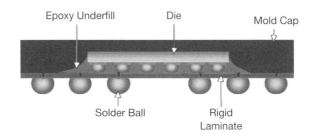

Figure 2.28. flip chip.

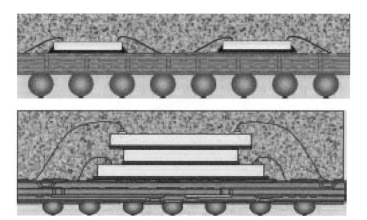

Figure 2.29. SiP, system in package.

CHAPTER 3

The Challenges in VLSI Circuit Design

18. WHAT IS THE ROLE OF FUNCTIONAL VERIFICATION IN THE IC DESIGN PROCESS?

IC design is the process of building miniaturized electronic components (transistors, capacitors, resistors, diodes, and inductors) on a monolithic semiconductor substrate by photolithography. When properly interconnected, these components can form complicated electrical networks for achieving desired functions. In general, IC design is divided into the categories of digital and analog design. Digital design produces components such as microprocessors, memories, and digital logics.

In digital designs, the main focus is the logical correctness of a design, along with its density, speed, and power usage. Analog designs, on the other hand, are more concerned with the physics of the devices, such as the gain, matching, power dissipation, and resistance. Analog design typically refers to the design of such components as op-amps, linear regulators, phase-locked loops, oscillators, and active filters. Fidelity of the analog signal amplification and filtering is critical.

The term *functional verification* refers to the verification of digital designs only. Analog designs have their own verification approaches, which are significantly different from what is discussed here. Modern ICs are enormously complicated, leading to the extensive use of automated design tools in the IC design process. This is especially true for large SoC designs (which are digital in nature).

Among the many aspects of digital IC design, one of the crucial tasks is functional verification. It is a process of proving that the chip we designed can faithfully perform the functions defined in production specification. Or, seen from another angle, it is a process of finding design problems, or functional bugs, which are introduced unintentionally during the chip design process. The two types of questions that functional verification asks are:

1. Does a design operate correctly?
2. Are two models of design under verification logically equivalent?

Functional verification only deals with logical and sequential properties of the design; it ignores the timing, layout, power, and manufacturing considerations. Verification involves three groups of specialists: architects who define what is intended, design engineers who implement the circuit to perform what is intended, and verification engineers who confirm that the circuit can perform what is intended. The functional verification of large-scale digital designs can be a very difficult task. It can be an *NP-hard* (nondeterministic, polynomial-time hard, a term from computational complexity theory) problem, for which no solution can be found that works well in all cases. In most projects, functional verification takes the majority of time and effort.

The difficult problem of functional verification in digital designs can be attacked using several methods as presented below. None of them can be applied to all of the designs, but each is helpful in certain circumstances:

- *Logic simulation.* Simulate the logic using a simulator before the hardware is built.
- *Simulation acceleration.* Apply special-purpose hardware to aid the logic simulation process.
- *Emulation.* Build a version of system using programmable logic (or other means) to emulate the real hardware. It is usually orders of magnitude faster than simulation.
- *Formal verification.* Attempt to prove mathematically that certain requirements are met or that certain undesired behaviors cannot occur.
- *Lint.* Flag suspicious HDL language usage that may indicate design problems.
- *Coverage check.* Check the percentage of functions or HDL codes that have been executed by simulation (or other means).
- *Prototyping.* No system is fully verified until physically operating in a real application environment for a decent period of time. Although simulation and emulation offer the benefit of flexibility, extensive observability in an environment fully integrated with design tools through hardware prototyping bridges the gap between the real world and the simulation environment and reduces the time and effort spent in verification.

Among these, simulation-based verification is widely used to study the design. In this method, a stimulus is provided to exercise each line in the

HDL code. Then a test bench is built to functionally verify the design by providing meaningful scenarios to check that, given certain inputs, the design performs to the specification. A simulation environment is typically composed of several components, as shown in Figure 3.1.

The purpose of the generator is to generate input vectors (stimuli) to activate the system under study. Modern generators can generate valid, biased, and random stimuli. In certain cases, randomness is necessary to achieve a high distribution over the huge space of the available input stimuli. In this approach, the user intentionally underspecifies the requirements for the tests. The generator randomly fills the gap. This mechanism allows the generator to create inputs that reveal the bugs not being searched for directly by the user. Generators can also bias the stimuli toward specific design corner cases to further stress the logic.

The driver translates the stimuli produced by the generator into the actual inputs for the design under verification. The generator may create inputs at a high level of abstraction, such as at the transactions level. The driver converts this type of input into actual design inputs as defined in the design interface's specification.

The simulator, which reads the design description in the format of HDL code or design netlist, produces the outputs of the design based on the design's current state (the state of the internal flip-flops) and the injected inputs.

The monitor converts the state of the design and its outputs to a transaction abstraction level for storage in a scoreboard database, which can be checked later on.

The checker validates that the contents of the scoreboard are legal. In some cases, in addition to the input stimuli, the generator also creates expected results. In these cases, the checker must validate that the actual results match the expected ones.

The arbitration manager manages these components to make them work together seamlessly.

At the end of verification process, different coverage metrics are defined and measured to assess the completeness of the verification process. This

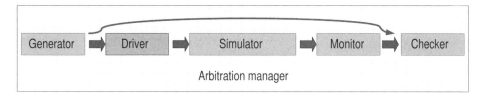

Figure 3.1. The simulation environment.

includes function coverage to check if every function has been exercised, statement coverage to check if each line of HDL code has been exercised, and branch coverage to check if each direction of every branch has been exercised. None of these metrics is sufficient to prove that a design will certainly work, but they are helpful in pointing out the areas of HDL code that have not been tested.

In practice, the task of verification must be carried out throughout chip development: during system design stage, during the logic design stage, after logic synthesis, and after the place and route stage. Verification can be performed at different levels during these stages: system level, RTL level, gate level, and even transistor level.

As the size of VLSI design projects increase and integration levels deepen, chips will function more and more like systems. Consequently, the verification task is divided into two categories: IP standalone verification and SoC verification. In this divide-and-conquer approach, the IP verification focuses on verifying the IP block's behavior against its specification. The SoC verification checks the interconnections of various on-chip IPs. It also checks for unexpected interactions between IPs.

As the SoC integration level increases, functional verification at the system level is drawing more attention and *Electronic System Language (ESL)* will definitely emerge as a standard tool in the near future, which may help ease some of the pain in architecture exploration and system-level verification.

In summary, verification is a very significant aspect of the chip development project. It is a never-ending process carried out until the chip is ready for production. In some cases, functional bugs are continually found even after the design is already being used in the field as a qualified product. More than one-half of the chip development effort in large SoC projects is often spent on verification.

19. WHAT ARE SOME OF THE DESIGN INTEGRITY ISSUES?

As design geometry shrinks and chip speed increases, electric signals inside the circuit experience a totally different environment than they did in the past. First of all, the circuit speed depends more on interconnecting wire delay than on logic gate delay. As process geometry gets smaller, the interconnecting wires correspondingly get closer, as shown in Figure 2.5 (see also Question 14 in Chapter 2). Thus, the cross coupling through the coupling capacitance between the wires has become more severe. This interference between different signal paths certainly degrades signal quality. For example, interference can make the device operate incorrectly, more slowly, or even fail completely; and it can create yield problems. This cross talk problem is the most crucial issue in the design integrity arena.

Electromigration (EM) is another design integrity issue. It is the unwanted transport of material caused by the gradual movement of the ions in a conductor (such as the copper and aluminum used in ICs) due to the momentum transferring between conducting electrons and diffusing metal atoms. Electromigration decreases the reliability of ICs. It leads to the eventual loss of one or more circuit connections and, consequently, the failure of the entire circuit. Because layout geometries are smaller now, the current densities inside wires are correspondingly higher. As a result, the practical significance of the electromigration problem increases.

For any chip to function correctly, the logic cells inside the chip must be provided with adequate power supply voltage. This task of power distribution over the chip is achieved by the power grid (made of metals) on the chip. When electrical current flows through a metal (which behaves as a resistor), it produces a voltage difference between the ends of this metal. This voltage difference is referred as the *IR* drop. The degree of *IR* drop over the power grid must be within a certain limit so that an acceptable voltage level can reach the cells on the chip. Otherwise, the chip performance degrades. Since there are often large currents present in the power grid geometries, the power busses are especially sensitive to *IR* drop, as well as to EM.

Another design integrity issue is associated with the gate oxide of a transistor and is referred as gate oxide integrity (GOI). A GOI check is a method of checking that none of the on-chip MOS transistor gates experience voltages higher than they are designed for, for extended periods. Such occurrences could damage the gate structure and cause the chip to fail.

Electrostatic discharge (ESD) also is a serious issue in solid-state electronics. It is the sudden and momentary electric current that occurs when an excess of electric charge finds a path from an object at one electrical potential to another object at a different potential. ESD events occur only among insulated objects that can store electric charge, not among conductors. Because transistors within IC chips are made from semiconductor materials of silicon and insulating materials such as silicon dioxide, they can suffer permanent damage when subjected to the high voltages of ESD events. Manufacturers and users of integrated circuits must take precautions to avoid this problem. During IC design and the implementation stage, special design techniques are employed so that device input and output pins, which are exposed to the outside world and subjected to ESD events, are not damaged from such events.

The problem of latch-up also falls into the design integrity category. It is the unintended creation of a low impedance path between the circuit's power supply rails. Such an occurrence of a low impedance path can trigger certain parasitic devices within the circuit structure, which then act as a short circuit and lead to failure. Worst of all, this large short-circuit current can lead to a circuit's destruction. Thus, during the design phase, a circuit is designed to be latch-up resistant. Layers of insulating oxides that surround both the NMOS and the PMOS transistors can break the parasitic structure between these transistors and thus prevent the latch-up.

As circuit speeds approach several hundreds MHz or even GHz, the inductance effect of the wires emerges as another design integrity problem.

These problems are the major issues in the design integrity arena. There are some other issues either not mentioned here or that will emerge as process technology continues to advance. These design integrity problems either cause the chip to malfunction immediately or impair the life span of the chip. They are among the reasons that make today's VLSI circuit design very challenging.

20. WHAT IS DESIGN FOR TESTABILITY?

During the chip development process, the designers must not only guarantee the chip's functionality, but they also must ensure its testability for volume production. The extra effort that designers incorporate into the development process for this purpose is called *design for testability,* or *DFT.*

Design for testability is an essential part of any production chip since the IC manufacturing process is inherently defective. If this extra circuitry of testability is not presented in the design, the chip manufacturer cannot confidently deliver the chip to its customer. If a bad part, or malfunctioning chip, is delivered to a customer and the customer uses this part to build a system and eventually sell this system to end users, the resultant financial damage could be significant. In volume production, if bad parts are accidentally delivered to a customer often and exceed certain levels, profits can vanish.

The action of design for testability is to add extra circuitry inside the chip so that the chip can be tested after the manufacturing process. If the chip does not behave as expected, it is scrapped. The financial damage is much smaller and controlled using this approach.

Design for testability targets problems introduced in the chip manufacturing process, as depicted in Figure 3.2. It is not intended for discovering functional bugs. In other words, it focuses on the chip's structural defects, not on logic flaws. Although design for testability cannot completely guar-

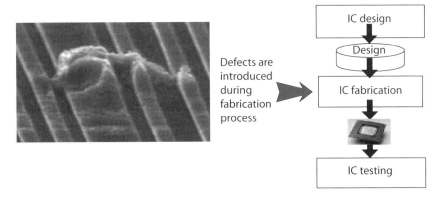

Figure 3.2. Design for testability.

antee the safety of the chips delivered to the customer, it can significantly diminish the *defect part per million (DPPM)* level.

Before the dies are cut from the wafer and sent to the *automated test equipment (ATE)* for individual testing, they are subjected to preliminary electrical and burn-in testing. Burn-in is a temperature/bias reliability stress test used in detecting and screening potential early life failures. This is called wafer-level testing and burn-in.

Wafer-level testing employs a wafer probe to supply the necessary electrical excitation to the die on the wafer through hundreds or thousands of ultra-thin probing needles that land on the bond pads, balls, or bumps on the die. During the wafer-level testing and burn-in, the electrical bias and excitation required by the devices are delivered directly to the interconnection points of each die on the wafer. The required die temperature elevation is achieved by the wafer probe through a built-in hot plate that heats up the wafer to the correct junction temperature. [In contrast, during the testing of individual ICs, electrical bias and excitation are provided by the ATE to the *device under test (DUT)* by mechanically contacting its leads.] In burn-in, the units are placed on burn-in boards, which in turn are placed inside burn-in ovens. The burn-in ovens provide the electrical bias and excitation needed by the devices through these burn-in boards.

Wafer-level testing and burn-in is a prescreen test. Only parts that pass this test undergo back-end processing (assembly and final test). They also provide additional information on identifying design and process problems. Figure 3.3 shows an example of the testing results from wafer-level testing. Among the dies on this wafer, the lighter boxes represent the good dies and the bad dies are the dark boxes. Such information is helpful during the design and process development phases for debugging purposes.

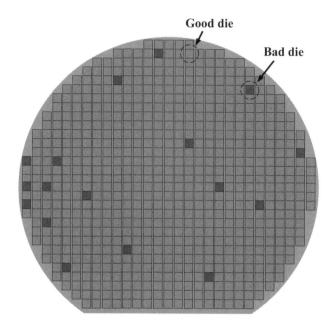

Figure 3.3. Test results of dies on a wafer.

21. WHY IS REDUCING THE CHIP'S POWER CONSUMPTION SO IMPORTANT?

Ideally, when designing a chip, we want the chip to use the least power possible without sacrificing chip performance. In general, better performance and higher speed require greater power consumption. Greater power consumption in turn demands more sophisticated cooling systems, more expensive packaging, and larger batteries if the chip is used for mobile applications. Therefore, one of the challenges in designing VLSI chips is to minimize the power usage by all means possible and make the chips environmentally friendly.

There are certain techniques for doing this but they are performed at the expense of other design merits such as area, speed, or design complexity. However, in some cases, smarter architecture at the system level can achieve the goal of significantly reducing the power usage without negatively impacting other design aspects.

This issue of cutting down a chip's power usage will become increasingly important as future devices are more power sensitive. It will become more of a problem, too, as leakage current gains influence on the overall power consumption of shrinking process geometries.

22. WHAT ARE SOME OF THE CHALLENGES IN CHIP PACKAGING?

As addressed in Chapter 2, Question 17, a chip package is a housing in which the chip resides for plugging into (socket mount) or soldering onto (surface mount) the printed circuit board. As we approach the scenario of having hundreds of millions of transistors, thousands of I/Os, and hundreds of watts of power on a chip, we face the challenge of how to package these monster chips. Among the challenges are how to minimize the signal distortion introduced by the package, how to package the chip economically (in the least expense way), how to conduct the heat out of the chip efficiently, and how to reduce the footprint of the package.

In addition to these challenges, there is the trend of chip and package codesign. This trend is driven by the fact that low-power, high-speed designs are becoming the mainstream of the modern chip design business. The power supply voltage for designs of 90 nm and below could be as low as 1 V. At this voltage level, the IR drop introduced by the package power plane must be factored into the chip design. Also, as the data rate gears up to tens of Gb/s (with rise and fall times of about 10 ps), the package exhibits complex behavior that can not be faithfully predicted by a simple, lumped-element circuit but requires an electromagnetic field solution. This package-related signal integrity impact on chip performance must be considered during chip design.

There is also an increasing interest in the multiple, stacked-die *system-in-package (SiP)* method as a realistic alternative to the SoC approach. This is also known as the *multichip module (MCM)*. As shown in Figure 3.4, a large system might require functional blocks of digital logic, processors, large amount of memory, and some analog functions. Traditionally, the system could be constructed using discrete components of individual packages. The system-on-chip (SoC) approach integrates the functional blocks in one chip. Thus, it could significantly reduce the overall system cost. However, to further trim down the cost, the SiP approach, which packs multiple dies into one package, could offer an attractive alternative. The chips are stacked vertically or placed horizontally alongside one another inside the package. They are internally connected by fine wires that are buried in the package or by using solder bumps to join the stacked chips together.

The potential cost saving comes from this fact: each functional block is designed and implemented in its appropriate technology. For example, a processor can be implemented in the most advanced digital process, whereas digital logic could be built on a process one or two generations behind. Memory would be constructed by a memory-oriented process that is three or

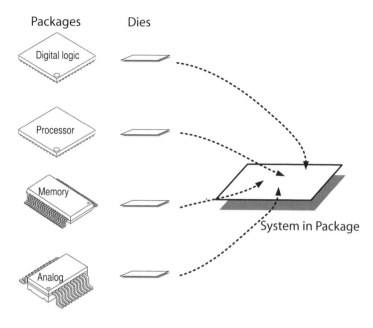

Figure 3.4. The system-in-package approach.

more metal layers less than the digital process. Analog functions could be designed in an analog-friendly process. Using this SiP approach, the overall system performance would improve since each function would be built with its own special process. Additionally, the test cost could be lower since dies with different functions are tested separately. And, finally, the yields could be higher since the different dies would be manufactured independently. However, there are trade-offs between SoC and SiP. In a real application environment, the optimum solution can only be achieved through detailed analysis, case by case.

23. WHAT ARE THE ADVANTAGES OF DESIGN REUSE?

As IC design practice gears toward higher performance, increasing complexity, and higher integration (higher density), the tools and methodologies that designers use in this practice struggle to keep pace. Managing the formidable design complexity is a major challenge in modern IC design. Design reuse is a strategy that helps bring the design effort back within reach.

Design reuse is the approach of using previously designed, verified, and even laid out blocks in a new design project. Its advantages include: shortening the design and verification cycle, shortening the physical design cycle in the case of layout-ready blocks, shortening the software development cycle, and reducing risk. Design reuse makes it faster and cheaper to design and build a new product since the reused components are not only already designed but also tested.

Reusability enables designers to build larger parts from smaller ones. In a reusable design, a function is common to or duplicated in several applications. Developers can reuse a component in similar or completely different applications. For example, a component designed as part of a general DSP processor unit can be reused in a handheld device or a set-top box. Reusability encourages designers to identify commonalities among their different applications and use these commonalities to build systems faster and cheaper.

As a methodology, design reuse involves the tasks of building, packaging, distributing, installing, configuring, deploying, maintaining, and upgrading reusable modules.

The widely adapted *intellectual property (IP)* approach is one example of design reuse. IPs are classified as soft, hard, or netlist. Soft IP is defined as a design block with only HDL description. Hard IP is the mask-layout-ready block, whereas netlist IP refers to a design block with a gate-level netlist but without layout. These different IP styles each have their own advantages and disadvantages. For example, soft IP is flexible; it can be migrated to new technology nodes without too much pain. Hard IP is usually performance or area optimized, but it is difficult to migrate.

The IP-enhanced design methodology has been used widely in chip design. In modern SoC chip designs, there is hardly a case without some type of IP incorporated. There are many strategically significant IPs available from various vendors, such as ARM processors (used mostly in embedded applications). IP reuse is a proven methodology that has improved design productivity notably.

Although design reuse is one of the most efficient ways of easing the time-to-market pressure, this approach is not without its challenges. The major problems are security and compatibility. The silicon vendors, IP providers, and EDA industry have already realized this problem and are now working together to create a standard for IP interchange.

24. WHAT IS HARDWARE/SOFTWARE CODESIGN?

Hardware/software codesign is defined as the simultaneous design of both hardware and software to implement a desired system. It is an interdisciplinary activity that brings concepts and ideas from different disciplines (such as system modeling, hardware design, and software design) together to create an integrated IC design system. As the leading feature of modern SoCs is the embedded microprocessor, which has both hardware and software contents, hardware/software codesign has moved from an emerging discipline to a mainstream technology. Hardware/software codesign also goes hand in hand with coverification, which is the simultaneous verification of both hardware and software.

The embedded microprocessors-based design practice separates the complex design problems into two subproblems: the design of the embedded microprocessors and the design of software that runs on the processors. Generally speaking, for an SoC-based system, software is used for features and flexibility, whereas hardware is used for performance. Without careful design considerations of the trade-offs between hardware and software, the resultant silicon could be too slow (not enough throughput), or too fast (more expensive than necessary). *Return on investment (ROI)* is only achieved when both hardware and software are working together effectively in the field.

Traditionally, the approach is to first develop the hardware platform and then write the software that runs on it. This method makes it easy for software engineers since they are working with a fixed execution engine and robust development tools. However, this sequential development of hardware and software leads to a prolonged development cycle that could miss the market window. Furthermore, the potential hardware errors that are detected during software work could force the development cycle to reset back. This bug fix in a later stage is very costly and could put projects in an unfavorable competitive position. Thus, concurrent hardware/software development is a requirement for modern SoC designs. Any chip vendor who can successfully achieve this goal will have a noticeable time-to-market advantage.

The design flow for the codesign approach could be shown in Figure 3.5. The codesign process starts with specifying system behavior. Then the system is divided into smaller subsystems of hardware and software. During the early stage, designers often strive to make everything fit in software and offload only crucial parts of the design to hardware to meet speed, power, size, and cost constraints. After the initial hardware and software partition, cost analysis is performed, which is based on estimations of hardware and software implementations. Hardware cost metrics are development time, chip area, power consumption, and testability. Software cost metrics include execution time and the amount of required program and data memory.

Following the cost estimation, hardware and software specifications are composed, the hardware is synthesized, and the software is compiled for the targeted processor. Finally, the hardware/software cosimulation is per-

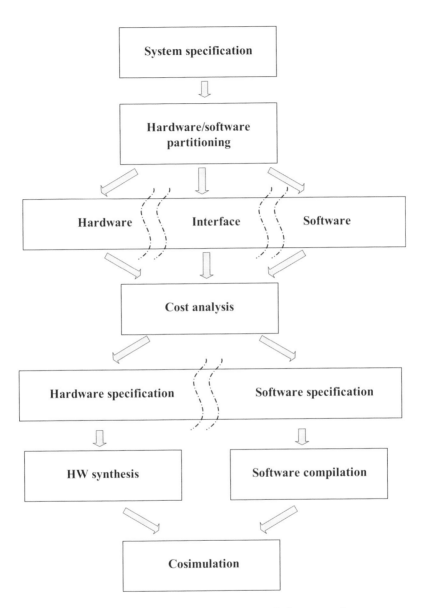

Figure 3.5. The hardware and software codesign approach.

formed. If the performance constraints are met and the cost of the design is acceptable, the codesign process stops. Otherwise, a repartitioning step is executed. During this repartitioning process, algorithms produce a number of different solutions in a brief expanse of computation time. This enables the designers to compare different design alternatives to find appropriate solutions for different objective functions (e.g., high performance, low cost, or low power). This optimization process is continuously performed until a sufficient system implementation is found. From this discussion, it is clear that a unified design methodology, or a system-level design language, that supports specification, covalidation, and cosynthesis of both hardware and software is the foundation of the hardware/software codesign approach.

As SoC systems become more complex with more embedded processors, the boundary between hardware and software increasingly blurs. For an efficient final implementation, engineers must have an appreciation of both the hardware and software aspects of a processor-based system design.

25. WHY IS THE CLOCK SO IMPORTANT?

A clock is an electric signal that oscillates between a high state and a low state. It is usually a square wave with a predetermined period (frequency), as shown in Figure 3.6. In synchronous digital circuits, the clock signals coordinate the actions of all the circuit components on a chip. The circuits become active at either the rising edge, the falling edge, or both edges of the clock signals for synchronization. The issue associated with clock signals is the most important design factor in any VLSI chip design.

Synchronization is a task in timekeeping that requires the coordination of events to operate a system in a harmonic fashion. In an electronic circuit in which millions of events occur every second, the synchronization of these events is the key to achieving the desired functions. During the process of synchronization, for some applications, relative timing offsets between events must be known or determined. For others, only the order of the events is significant.

The synchronous design principle can significantly simplify the implementation task in chip design. The design and verification burden are eased

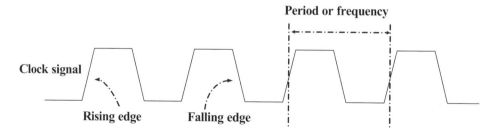

Figure 3.6. A clock signal waveform.

greatly. This is especially true for large SoC designs in which design complexity is dreadful. As an example, the synchronous design principle enables the technique of *static time analysis (STA)*, which is an essential tool for achieving timing closure. Synchronous design style also enables the method of formal verification, which is an important approach for logic verification. Without synchronous design principles, or clocks, it is impossible to construct the complicated SoC chips that we build today.

As just addressed, in today's VLSI chip design environment most integrated circuits of sufficient complexity require clock signals to synchronize different parts of the chip and to account for propagation delays. However, as chips get more complex and clock speeds approach the gigahertz range, the task of supplying accurate and synchronized clocks to all of the circuit components becomes more and more difficult. Furthermore, the voltage and current spikes associated with clock switching become harder to control because millions of components are switching at roughly the same time. As a result, the asynchronous self-clock circuit design approach has been explored with great interest.

Figure 3.7 shows the principal ideas of synchronous and asynchronous design styles. In the synchronous design approach, the actions are coordinated by the clock signal as data is moved from register to register. In contrast, in clockless asynchronous designs, the actions are coordinated by a handshake mechanism between the blocks. When a block must initiate a data transfer, it first sends a *request signal (REQ)*. The intended block issues an *acknowledge signal (ACK)* when it is ready for the transfer. All of the data communication inside the asynchronous block is accomplished though certain handshake mechanisms without using the clock. The advantage of

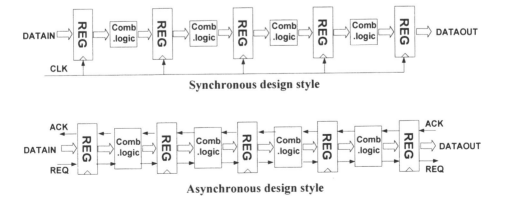

Figure 3.7. Synchronous and asynchronous design styles.

this method is that it eliminates the design overheads associated with clock structure. In some cases, the asynchronous design approach can potentially increases the data throughput as well. It also provides the superior modularity that is preferred for SoC designs. Due to the clockless feature, it is more robust against the process, temperature, and voltage variations in terms of wire delay. It definitely lessens the power supply noise by reducing the current peak around the clock edges. The overall power consumption is also trimmed since the clock-related power usage is now nonexistent.

However, the asynchronous design style cannot be easily implemented in large designs due to the lack of the corresponding EDA tools. Additionally, the testing of the asynchronous design is more difficult than that of the synchronous circuit. Currently, there is a design approach called *globally asynchronous locally synchronous (GALS)* that combines the advantages of both asynchronous and synchronous. Figure 3.8 presents this technique. In this configuration, certain low-level blocks are synchronously designed. Then asynchronous wrappers with handshake mechanisms are constructed around such blocks. At the chip level, communication is accomplished through asynchronous interconnection.

Clocks are also essential for certain types of analog circuits to function. For example, analog-to-digital converters and digital-to-analog converters all work on clock signals. The internal circuitry of these converters, and thus the signal conversion, are triggered by the clock edge.

In summary, for a VLSI chip to function, reliable clock signals must be provided one way or another.

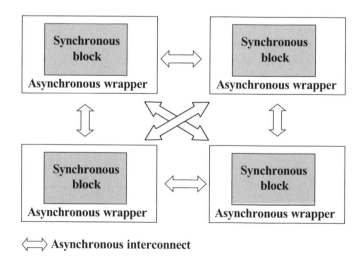

Figure 3.8. A globally asynchronous locally synchronous (GALS) system.

26. WHAT IS THE LEAKAGE CURRENT PROBLEM?

Power consumption is now the major technical problem facing the semiconductor industry. There are two principal sources of power dissipation in today's CMOS-based VLSI circuits: dynamic and static power. Dynamic power, which results from transistor switching and repeated charging and discharging of the capacitance on the outputs of millions of logic gates on chip, is the energy consumed by the chip to produce a useful outcome. Static power refers to the leakage current that leaks through transistors even when they are turned off. It is the power that is dissipated through transistors without producing any useful operation. Until very recently, only dynamic power has been a significant source of power consumption. However, as process geometries continuously shrink, smaller channel lengths have exacerbated the leakage problem. In particular, as process technology advances to the sub 0.1 μm regime, leakage power dissipation increases at a much faster rate than dynamic power. Consequently, it begins to dominate the power consumption equation.

For deep submicron MOSFET transistors, there are six short-channel leakage mechanisms, as illustrated in Figure 3.9. I1 is the reverse-bias p–n junction leakage. I2 is the subthreshold leakage or the weak inversion current across the device. I3 is the gate leakage or the tunneling current through the gate oxide insulation. I4 is the gate current due to hot-carrier injection. I5 is *gate-induced drain leakage (GIDL)*. I6 is the channel punchthrough current. Among these currents, I2, I5, and I6 are off-state leakage mecha-

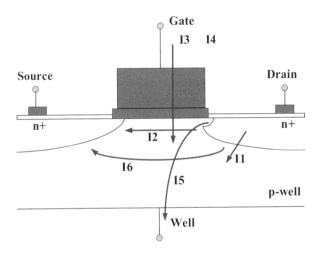

Figure 3.9. The leakage current mechanisms in MOSFET transistors.

nisms since they only exist when the transistor is in off state. I1, I3, and I4 can occur on both on and off states. The leakage currents are influenced by threshold voltage, channel dimensions (physical), channel/surface doping profile, drain/source junction depth, gate oxide thickness, and V_{DD} voltage. Currently, the two principal components of static power consumption are the subthreshold leakage I2 and gate leakage I3.

Most of the operations of modern VLSI chips can be classified into the two modes: *active* and *standby*. During the active mode of circuit operation, the total power dissipation includes both the dynamic and static portions. While in standby mode, the power dissipation is due only to the standby leakage current. Dynamic power dissipation consists of two components. One is the switching power due to the charging and discharging of load capacitance. The other is short-circuit power due to the nonzero rise and fall time of input waveforms. The static power of a CMOS circuit is only determined by the leakage current through each transistor. In other words, dynamic power is related to the circuit switching activity. In contrast, static power is proportional to the total number of transistors in the circuit regardless of their switching activities.

In general, dynamic power dissipation is expressed as $P_{dyn} = \alpha f C V_{DD}^2$, where α is the circuit switching activity, f is the operation frequency, C is the load capacitance, and V_{DD} is the supply voltage. In the past several decades, as CMOS devices scaled down, supply voltage V_{DD} has also been trimmed down to keep the power consumption under control (since the power usage has quadratic dependence on V_{DD}, according to the equation). Accordingly, the transistor threshold voltage has to be commensurately scaled to maintain a high drive current and achieve performance improvement. However, this threshold voltage scaling also results in a substantial increase in subthreshold leakage current. Consequently, leakage power becomes a significant component of the total power consumption in both the active and standby modes of operation. Figure 3.10 shows the dynamic and static power dissipation trend in the foreseeable future, based on the International Technology Roadmap for Semiconductors (ITRS) 2002 projection. As shown, at some time in near future the static power dissipation will inevitably passes dynamic power as the dominating factor in total chip power consumption.

To suppress the power consumption in deep-submicrometer circuits, it is necessary to reduce the leakage power in both the active and standby modes of operation. The reduction in leakage current can be achieved using both process-level and circuit-level techniques. At the process level, leakage reduction can be achieved by controlling the dimensions (length, oxide thickness, and junction depth) and doping profile in transistors. At the circuit lev-

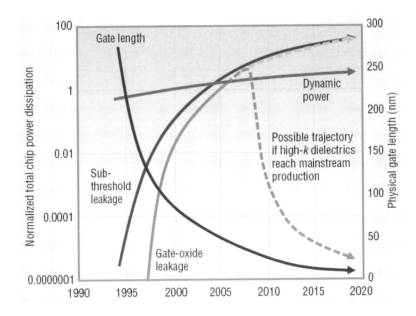

Figure 3.10. The dynamic and static power dissipation trend, ITRS projection.

el, threshold voltage and the leakage current of transistors can be effectively controlled by controlling the voltages of different device terminals (drain, source, gate, and body or substrate). In practice, several circuit design techniques have been suggested for leakage reduction in digital circuits (logic and memory), such as transistor stacking, multiple V_T, dynamic V_T, and supply voltage scaling (multiple and dynamic V_{DD}).

In system-level implementation, there are two complementary approaches to limiting leakage current: *statically selected slow transistors (SSST)* and *dynamically deactivated fast transistors (DDFT)*. The static approach is design independent and may be implemented with multiple-threshold (multiple-V_T) libraries and associated design tools that support these libraries. Most foundries today offer multiple-V_T libraries for processes of 0.13 μm and below that contain both fast cells (high leakage, low V_T) and logically equivalent slow cells (low leakage, high V_T). The dynamic approach requires that the chip designer employ techniques to dynamically deactivate parts of the chip during periods of inactivity. This control mechanism must be built into the system during the design process, and DDFT is thus design dependent.

If care is not taken, leakage power can lead to dramatic effects that may cause a circuit to fail to function properly. Large leakage current can increase the standby power dissipation to an unacceptable level or it can lead to excessive heating, which consequently requires complicated and expensive cooling and packaging techniques. In the beginning, only leading edge microprocessors were affected by this leakage current problem, but now leakage current has become a critical design parameter for all the nanometer chips.

In summary, for modern VLSI chip design, the issue of controlling leakage current has moved from backstage to center stage.

27. WHAT IS DESIGN FOR MANUFACTURABILITY?

Although there is no industry-wide consensus, the term *design for manufacturability* is roughly described as the specific work of analysis, prevention, correction, and verification that targets improving a product's yield. It is different than the post-GDSII resolution enhancement techniques, such as *optical proximity correction (OPC)* and *phase shift masking (PSM)*. The keyword in this term of design for manufacturability is *design,* which refers to the work performed during, not after, the design phase for this purpose. Design for Manufacturability is often interchangeable with the term *design for yield.*

As process technology shrinks below 0.13 μm, the issue of design for manufacturability has emerged as a serious challenge of reaching the goal of acceptable manufacturing yield. This is caused by the fact that the IC features are now smaller than the wavelength of the light that creates them. As a result, the layout patterns generated in the design phase cannot be reproduced faithfully. This scenario is analogous to printing a thin line with a wide paintbrush. Figure 3.11 demonstrates one such problem. In relatively old process nodes, such as 0.25 μm or 0.18 μm technologies, engineers paid attention to issues such as wide metal slotting, dummy metal fill for density requirement, redundant vias, and so on. These issues also fall into the category of design for manufacturability.

In the past, most ASIC design engineers have been isolated from the fabrication process. The design and manufacturing worlds have been treated as two separated entities, connected only by design rules and sometimes additional recommended rules. IC designers could safely assume successful fabrication of their chips as long as these chips rigorously met the rules. Any yield-related problem was considered the foundry's responsibility. However, in the case of today's deep submicron technologies, this no longer reflects the underlying physics of the manufacturing process. Even if the chips are violation free according to the rules provided by the foundry, they can still suffer significant yield loss.

Phase shift mask (PSM) and optical process correction (OPC) are examples of procedures used to obtain better pattern printability. They are post-

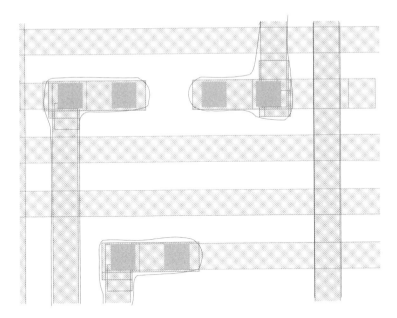

Figure 3.11. The layout geometries are not faithfully reproduced.

design steps based on layout *Gerber data stream information interchange (GDSII)* data, a database format for exchanging layout geometries information. The problem with these techniques is that they are performed after the layout, which is too late in the design flow. If the quality of the initial layout is poor, it is too costly to address the DFM problems in term of area, timing, and schedule. And, in some cases, it is simply impossible to satisfy the DFM-required fixes.

In current design flows, the place and route tool does not have the capability to account for complex lithographic interactions and effects. However, to address the layout geometry printability problem, there is a need for such DFM-aware, place and route engines. Such engines must be embedded with the printability analysis capability (using knowledge of the requirements and limitations of the downstream OPC process) right inside the engine. In this way, layout patterns with DFM issues can be avoided. In other words, when manufacturing-unfriendly spots are identified in the layout, the tool directs the layout engineer or the place and route engine to make appropriate corrections. It also identifies the locations where extra spacing must be added for downstream OPC.

The issue of DFM is closely tied to yield modeling and analysis. This study is both theoretical in nature and process specific. In theory, DFM-related process simulations use mathematical formulas that are based on physics and parameters extracted from the process line. In practice, the real yield curves of DFM methodology are very difficult to generate since they require a large amount of data, and some of these data are very financially sensitive. Nevertheless, a cost model should be incorporated in the DFM tools to justify their usefulness.

In summary, design for manufacturability covers the extra work that targets product yield, which has a direct impact on profit margin.

28. WHAT IS CHIP RELIABILITY?

A VLSI chip not only needs to function correctly during the first few days, weeks, or months, but also has to function reliably for its entire life span. The life span of any chip designed for commercial use is usually defined as 100,000 hours or 11.4 years. However, during the design, fabrication, assembly, and test of the IC many factors can contribute to its early failure. This aspect of chip development is referred as chip reliability. The difference between test-related failures and reliability-induced failures is that test-related failures are detectable prior to product shipment, whereas reliability-induced failures occur after shipment. Reliability failures are those that are physically undetectable with present test technology. They occur during the actual usage of the chip.

The key environmental agents that can affect chip reliability are voltage, current, temperature, and humidity. Transistor gate oxide breakdown (GOI), hot carrier stress (HCS), negative bias temperature instability (NBTI) of PMOS devices, stress-induced voiding (SIV), metal damage caused by electromigration, and the breakdown of the intermetal dielectric are often the physical mechanisms behind reliability failures.

Typically, IC reliability is represented by a bathtub curve shown in Figure 3.12, which shows the failure rate of ICs with respect to time. This curve has three individual regions: early life failure, useful life, and wear out. Each of these regions has its own potential failure mechanism. Early life failures, also called infant mortalities, are typically caused by defects and contaminants introduced during the manufacturing process. With today's well-controlled fabrication and assembly processes, very few early life failures occur. However, the materials that make up the gate and capacitor oxides, contacts, vias, and metal lines in the fabrication process can wear out over time with the application of constant voltage and current. This effect is cumulative and can eventually lead to opens and shorts in the circuit or change the electrical characteristics required for the product to function accurately. This is the failure mechanism in the wear-out region. It indicates the end of the chip's useful life.

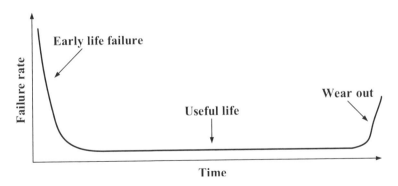

Figure 3.12. Product reliability.

The failure rate in the useful life region is low and near constant, which implies low-level residual defects. Failures in this region are also due to electrical overstress or ESD events. This type of damage can occur when the ICs are handled or transported without the use of ESD protection such as ground straps; ESD-resistant trays, tubes, or reels; or properly grounded machines in assembly. Protection from ESD must be part of the circuit design consideration and manufacturing/assembling environment. Ideally, the shape of the bathtub curve should have a brief region of early life failure and a very long region of useful life.

Chip reliability is a complicated issue that involves both the work of the chip designer and the quality of the fabrication at foundry. The foundries are strongly motivated to mitigate any physical mechanism that may cause a chip to fail. They need to understand the physics of every failure mechanism and identify wafer processing steps that may detrimentally influence the mechanisms. Once the wafer processing steps are optimized for maximum chip lifetime, the foundries develop design rules intended to prevent chip designers from overstressing the devices and causing the expected lifetime to fall below the target. The design rules are embodied in the form of maximum operating voltage, transistor channel length constraints under certain bias conditions, maximum current per unit line width in the metal interconnect, maximum current per via or contact, and certain constraints upon very wide metal lines. Failure to comply with these reliability design rules can lead to shorter chip lifetime.

The semiconductor industry is approaching physical, material, and economic limits as aggressive scaling continues. This results in formidable reliability challenges. Some of the emerging reliability challenges include increased gate leakage currents as oxides become so thin that direct tunneling occurs between the channel and the gate, the trade-off of reduced reliability safety margins for increased performance, the need for improved design-for-reliability capability, the need to address the chip's increased sensitivity to background radiation that results in increased single-event logic state upset probability, and the need to address burn-in as more products are used at near burn-in temperatures.

VLSI circuits designed for environments with high levels of ionizing radiation have special design-for-reliability challenges. A single charged particle of radiation can knock thousands of electrons loose, causing electronic noise, signal spikes, and, finally, inaccurate circuit operation, especially in the case of digital circuits. This is a particularly serious problem in the circuits being used in the artificial satellites, spacecraft, military aircraft, nuclear power stations, and nuclear weapons. To ensure the proper operation of such systems, manufacturers of these circuits must employ various methods of radiation hardening. The resultant systems are said to be radiation-hardened, or RADHARD for short.

There are two approaches on designing RADHARD circuits. One is at the physical or process level, the other is at the circuit or system level. At the physical or process level, techniques include:

- Manufacturing the circuits on insulating substrates rather than the usual semiconductor wafers. Silicon oxide (SOI) and sapphire (SOS) are commonly used. Whereas a normal commercial grade chip can withstand between 5-10 krad of radiation, space-grade SOI and SOS chips can survive doses many orders of magnitude greater.
- Shielding the package against radioactivity to reduce the exposure of the bare device.
- Replacing capacitor-based DRAM with more rugged SRAM.
- Choosing a substrate with a wide band gap (silicon carbide or gallium nitride) that has a higher tolerance to deep-level defects.
- Using depleted boron (consisting only of isotope boron-11) in the borophosphosilicate glass layer protecting the chips, as boron-10 readily captures neutrons and undergoes alpha decay.

The methods used in circuit- and system-level RADHARD circuits include:

- Error-correcting memory, which has additional parity bits to check for, and possibly correct, corrupted data.
- Redundant elements at the system level. For example, several separate microprocessor boards may independently compute an answer to a

calculation; their answers are compared. Any system that produces a minority result is required to recalculate. Logic could be added such that, for repeated errors in the same system, the board would be shut down.

- Redundant elements at the circuit level as well. A single bit may be replaced with three bits and separate voting logic for each bit to continuously determine its result. This increases chip area. But it has the advantage of being fail-safe in real time. In the event of a single-bit failure, the voting logic continues to produce the correct result.

- A watchdog timer hard reset as the last resort to other methods of radiation hardening. During normal operation, software schedules a Write to the watchdog timer at regular intervals to prevent the timer from running out. If radiation causes the processor to operate incorrectly, it is unlikely the software will work correctly enough to clear the watchdog timer. The watchdog eventually times out and forces a hard reset to the system.

In general, most radiation-hardened chips are based on their more mundane commercial equivalents with some manufacturing and design variations that reduce susceptibility to radiation and electromagnetic interference. Typically, the hardened variants lag behind the cutting-edge commercial products by several technology generations because it takes extensive development and testing to produce radiation-tolerant designs.

29. WHAT IS ANALOG INTEGRATION IN THE DIGITAL ENVIRONMENT?

Analog circuits are those circuits that monitor, condition, amplify, or transform continuous signals associated with certain physical properties, such as temperature, pressure, weight, light, sound, and speed. Analog circuits play a major role in bridging the gap between real-world phenomena and electronic systems.

Typically, most SoC chips are digitally oriented. If certain analog signal processing work is required, then the corresponding analog circuit must be integrated into the chip. For instance, in a SoC targeting a HDTV application, analog output signals are required to drive the analog CRT monitor. Analog video or graphic input signals must also be converted to digital format to be processed by the on-chip DSP engine. In such cases, analog components of high-speed analog-to-digital converters (ADCs) and digital-to-analog converters (DACs) are required as part of the chip. Other examples are high-speed serial links such as SErializer and DESerializer (SEDES). If this function is required for an SoC, then high-quality analog signal conditioning circuits must be present in the chip as well. Moreover, in communication-oriented SoCs, radio frequency (RF) circuits are an indispensable part of the chip. Furthermore, almost every SoC has one or more phase lock loops (PLLs) on board as timing sources. The PLL is a very analog-intensive component as well.

Compared with the approach of using separated chips for analog functions, integrating them into the SoC significantly reduces the overall system cost. However, this approach is not without its own problems. The key challenge is analog performance. This is because SoCs are primarily digital oriented and are on digital processes. Thus, on-chip analog circuits cannot be optimized as well as in an analog-friendly process.

The principal difficulty in using a digital CMOS technology for analog design is that the digital process is only optimized and characterized for the simple trade-offs among speed, area, and power dissipation. By contrast, analog circuits entail a multidimensional design space: noise, speed, voltage swings, supply voltage, gain, linearity, power dissipation, and input/output impedance, where almost every two parameters trade with each other. Compared with a digital circuit, which is only sensitive to timing variation, an analog circuit is additionally subjected to voltage level variation. Consequently, the design complexity associated with an analog circuit is much greater than with its digital counterpart. The device and circuit properties of interest in analog design include DC behavior, AC behavior, linearity, device matching, temperature dependence, and noise.

Furthermore, since the manufacturing process itself is not completely predictable, designers must account for its statistical nature. This is especially true for analog designs. Unlike board-level analog circuit designs, which permit the designers to select devices that have each been tested and characterized completely, the device values for an IC can vary widely; the designers cannot control these. The underlying cause of this variability is that semiconductor devices are highly sensitive to uncontrollable random variances in the manufacturing process. Uneven doping levels or slight changes to the amount of diffusion time can have significant effects on device properties. Consequently, analog IC circuits must be designed in such way that the absolute values of the devices are less critical than the identical matching of the devices.

To cope with the inherent variability of the individual devices built on chips, special design techniques are needed for analog IC designs, such as using devices with matched geometrical shapes so that they have matched variations, making devices large so that statistical variations become an insignificant fraction of the overall device properties, using the ratios of resistors (which do match closely) rather than the absolute resistors' values, segmenting large devices into parts and interweaving them to cancel variations, and using common centroid device layouts to cancel variations.

For digital processes, the variety of available active and passive devices is often limited and the devices are only characterized and modeled according to simple benchmarks, such as current drive and gate delay. Such a shortage of appropriate analog devices and the lack of analog characterization in the digital process can make the implementation of analog functions on a chip very challenging.

Another major issue is noise isolation. Typically, SoCs contain large digital contents with complicated clock structures. When their digital circuits toggle at the moment of clock switching, a lot of noise is generated. The on-chip analog circuits must be guarded from this noise. The power supply voltage distortion caused by digital switching must be controlled within certain levels for both the analog and digital components to work harmoniously in the same chip.

The integration of the RF function in SoC environment further complicates the issue. The performance issue of RF is multidimensional owing to the different requirements for the various RF building blocks: low-noise amplifier (LNA), mixer, oscillator, and power amplifier. The implementation of a highly integrated radio transceiver on a CMOS digital process is one of the most difficult challenges in the area of SoC integration today.

As a trend, more and more analog circuits will be found in future SoCs. Technically, there are many issues in the integration of analog and digital circuits. The know-how of this integration will be the key differentiating factor among competing companies.

30. WHAT IS THE ROLE OF EDA TOOLS IN IC DESIGN?

During the very early years of IC design, the chips were built by manually laying out every transistor of the circuit on a drawing board. It is unimaginable how many man-years would be required to design modern SoCs in this outdated way. It is the electronic design automation (EDA) tools that fundamentally changed the IC design and made today's multimillion gate designs possible.

In today's chip design environment, there are many EDA tools to help designers perform their work. Each of them targets a specific application.

The synthesis tool raises the design abstraction level from device/transistor to RTL, which is the single most significant factor that makes modern SoC design feasible.

The most commonly used EDA tools in today's IC design environments include:

- Simulation tools at the transistor level, switch level, gate level, RTL level, and system level.
- Synthesis tools that translate and map the digital RTL code to real library cells.
- Place and route tools, which perform the automatic layout based on various design constraints.
- Logic verification tools, which include formal verification tools and simulation tools.
- Time verification tools, which verify the design's timing quality.
- Physical verification tools, which verify the design's layout against manufacturing rules.
- Design for testability tools, which integrate testability into design and generate test patterns.
- Power analysis tools, which perform power dissipation analysis and IR drop analysis.
- Design integrity tools, which check a design's reliability-related issues, such as ESD, latch-up, EM, GOI, and antenna.
- Extraction tools, which extract the design's parasitics for back annotation.
- Rule checkers for checking the design's logical and electrical compliance with corresponding rules.
- Package design and analysis tools.

There are some other special tools, such as schematic capture tools for analog designers, layout tools for layout engineers, and process simulators for process and device engineers.

As the SoC integration level rises and chip size increases, the requirements for EDA tools have been pushed in the directions of *faster* and *larger*. In other words, to perform a specific task on a large SoC design, the corresponding EDA tool must have the capability of handling the necessary data as one integral part (without separating it into smaller pieces) and finish the task within a reasonable time schedule. With continuous innovations from the EDA industry and aided by ever-improving computing hardware, EDA tools have kept pace with the design complexity explosion reasonably well.

In summary, EDA tools make up the foundation of today's IC development activities. By utilizing these tools, engineers create miraculous wonders that are changing our world.

31. WHAT IS THE ROLE OF THE EMBEDDED PROCESSOR IN THE SOC ENVIRONMENT?

As addressed in Chapter 1, Question 8, embedded microprocessors are the brains of the SoC. A system-on-chip is built primarily around the processors. The key difference between the embedded processor and the general-purpose processor is that embedded processors are used in an environment that is surrounded by many other on-chip components, whereas the general-purpose processor is a standalone chip. As its name suggests, the general-purpose processor is designed for general usage. In contrast, the embedded processor is designed for a specific application. The performance of the general-purpose processor can be improved at the expense of power usage and silicon area. In other words, for the general-purpose processor, performance is the highest priority. However, for the embedded processor, cost and power consumption are more significant. A low-power processor is especially attractive to the SoCs used for mobile applications.

Several approaches have been employed to improve the performance of embedded-processor-based systems. The most commonly used increases clock frequency. However, this can result in increased power consumption. Additionally, since the performance of the external memory has not kept pace with processor technology, this processor and memory mismatch gap limits the overall system performance gain from the clock speed increase. Another approach is a multicore system with several processor cores on-chip to improve performance. But this is companied by the expense of larger areas and higher power usages. Multiissue processors with two or more execution units offer another alternative. But they also have a large-area penalty. Also, the software must be revised to make best use of the multiple pipelines. The multithreading approach, which supports multiple threads of software on a single core, provides some balance of the trade-offs. This approach obtains its performance gain by keeping the processor hardware as busy as possible.

Another direction in embedded processor development is the configurable core. It enables SoC designers to create silicon that is optimized to the end application and gives designers the freedom to retain necessary functionality while removing unneeded features. This produces an optimal balance of speed, area, and power for a specific application. The configurable core can also have extendibility so that SoC designers can achieve further gain in application efficiency by defining custom extensions to accelerate critical code execution.

Typically, embedded processors are delivered to semiconductor customers as IPs by IP vendors. Currently, the most popular processor platforms used for embedded SoC applications are the ARM and MIPS cores. In the configurable core market, the ARC core is most significant.

An issue closely related to the on-chip processor is embedded memory, which is critical for SoC software development. During the past several decades, the processor's performance has been improved greatly. However, memory performance has not caught up with the pace. As this performance gap widens, chip designers have placed greater emphasis on the development of embedded memory devices. The advantages of using embedded memories are as follows: interchip communication is eliminated; response time is faster, chip pin count is reduced, the number of chips at the system level is reduced so less board space is required, on-chip multiport memories can be easily realized, and, finally, memory capacity is specific for an application, resulting in reduced power consumption and greater cost effectiveness at the system level.

The main disadvantages of embedded memories follow:

- Size. They are generally larger in size (compared to standalone memory, in area per bit).
- Complexity. They are more complex to design and manufacture.
- Design and technology trade-offs. Because the optimized technology for a memory cell is often not the same as that for on-chip logic devices, there are trade-offs between design and technology.
- Processing. Processing becomes more complex as designers integrate different types of memory on the same chip.

Currently, embedded SRAM is widely used in SoC designs due to its easy integration with logic devices. Embedded DRAM is not as popular due to the complexity of DRAM process technology. Embedded DRAM capacitors that store data require several processing steps that are not needed when making logic devices. Also, the threshold voltage of DRAM transistors must be high enough to ensure that they do not cause memory cell capacitor leakage. This constraint on low subthreshold current may result in a speed penalty on the logic portion of the device. Until recently, DRAM has been the least-used embedded memory technology. However, it may become a more widespread solution as on-chip memory demand increases.

The nonvolatile embedded memory options include embedded ROM, embedded EPROM, embedded EEPROM, and embedded flash memory. Their reprogrammability and in-circuit programming capability provide highly flexible solutions to rapidly changing market demands.

CHAPTER 4

Cell-Based ASIC Design Methodology

32. WHAT ARE THE MAJOR TASKS AND PERSONNEL REQUIRED IN A CHIP DESIGN PROJECT?

An ASIC chip project typically starts with marketing research, which is followed by product definition and system-level analysis. After the system-level validation, the ASIC design flow carries out the chip implementation process, which turns the paper design into real hardware.

ASIC flow itself usually begins with RTL coding and functional verification. Next, the tasks of logic synthesis and place and route are carried out. It ends with final logic verification, timing verification, physical verification, and tapeout. Although it seems simple, chip implementation is actually very complicated. It can require numerous iterations among the various steps before final implementation is acceptable. Also, tremendous resources are required: license fees on commercial computer-aided design (CAD) tools, which are often very expensive; hardware, such as powerful CPUs with large memory capacities and disk space, is costly; and, most importantly, talented and dedicated individuals, who require competitive compensation.

The engineers needed for a chip project include:

- System engineers who define the chip at the system level.
- IC design engineers who compose the RTL codes for the digital blocks and design the circuits for analog components.
- Verification engineers who verify the functionality at both the block and chip level.
- Design-for-testability engineers who ensure that the chip is testable for volume production.
- IC implementation engineers to turn the design from a paper plan into real hardware.
- Software engineers to make the bare silicon chip into a useful electronic device.

- Application engineers to build the reference design for customers.
- Test engineers to write testing programs for production tests.
- Product engineers are needed during the chip's volume production to coordinate the operations between the design and manufacturing facilities and, generally, to ensure the smooth flow of chip production.

In most cases, a project (or program) manager is also assigned to an ASIC chip project to coordinate the design, test, production, and marketing activities of the entire project. S/he creates and tracks project milestones to keep the project on schedule and within budget. If necessary s/he also consults with business managers to adjust the milestones based on market conditions and design status and sometimes to acquire additional resources.

One of the key figures in a chip project is the design leader. The role of the design leader is to lead the design team from a technical, not a business or administrative, perspective. The design leader is not obligated to know, or be expert at, every detail of the technical aspects of building the chip. However, s/he must have a solid understanding of the major aspects of IC design. An underqualified design leader is a guarantee of disaster. This is simply because IC design is complicated. A single incorrect technical decision could result in a dreadful financial or schedule penalty, or even the disaster of missing the market window completely.

33. WHAT ARE THE MAJOR STEPS IN ASIC CHIP CONSTRUCTION?

As mentioned in Question 32, the major steps in ASIC chip construction are RTL coding, function verification, logic synthesis, place and route, final logic verification, timing verification, physical verification, and tapeout.

RTL coding is the step of translating the design intention, which is described in plain language such as English or Chinese, into a simulative computer language. In this way, simulation can be performed to verify design intention. Additionally, it must be possible for the RTL code to be turned into hardware successfully in later stages. In other words, the RTL code must be synthesizable. The two major languages used in this field are Verilog and VHDL.

Function verification is the process of verifying that the RTL code, created in either Verilog or VHDL, can perform the intended functions defined in the product specification. The main approach used in this process is simulation, with the aid of test benches that describe the system's operations (see Chapter 3, Question 18). A large number of test benches must be created in a typical project to cover as many chip operation scenarios as possible. The aim of function verification is to find design problems in RTL code or to find *bugs* as they are commonly called. Ideally, the number of bugs should be zero before the flow, progresses to the next step.

The step of logic synthesis turns the verified RTL code into real circuit hardware by using sophisticated algorithms. The main task involved in this process is mapping the logic functions defined in the RTL code to standard cells in a selected ASIC library, where each standard cell has a predefined function. The result of this step is a netlist, which has the instantiations of the cells used and a description of their interconnections.

Place and route is the process of laying out those cells and interconnecting wires in an automatic way. It is impossible to accomplish the layout process manually in any design of sufficient complexity.

After place and route, it must be functionally verified that the resultant physical entity can still achieve the design functionality defined in the RTL code. This physical entity must also be checked against design rules for manufacturability. The timing aspect of the design is examined at this stage as well. The final step is tapeout, in which the design data is generated in a certain format and sent to a manufacturing facility for fabrication.

34. WHAT IS THE ASIC DESIGN FLOW?

In today's world of VSLI circuit design and manufacture, as system architects integrate more and more system functions into one chip, IC design engineers must confront the difficult challenge of building giant entities of hundreds of millions of transistors. Two crucial requirements in this chip-creation process are build it correctly and build it fast. Nowadays, it is commonly agreed that it will cost millions of dollars to develop a multi-million-gate ASIC, from concept to silicon. The financial penalty of building something incorrectly and remaking it all over again is intolerable. Furthermore, in addition to being built correctly (performing as specified), it must also be built in a timely fashion. Otherwise, the market window for the product may be missed. There are too many examples of technically great products not being able to earn a dollar for their investors, simply due to the delay in their chip-creation execution. Fortunately, there is something called *design flow,* which IC implementation engineers can rely on in this very demanding business. A well-tuned design flow can help designers go through the chip-creation process relatively smoothly and with a decent chance of error-free implementation. And, a skillful IC implementation engineer can use the design flow creatively to shorten the design cycle, resulting in a higher likelihood that the product will catch the market window.

In principle, a design flow is a sequence of operations that transform the IC designers' intention (usually represented in RTL format) into layout GDSII data. In practice, a design flow is a sequence of executions that perform each individual task described previously. In composition, a design flow is a suite of software programs; they are either commercial CAD point tools or programs or scripts developed in-house.

A design flow, as a whole, is a wrapper that glues many software programs together. The execution sequence of these softwares is arranged according to the physics of hardware creation. The detailed tasks are:

- Logic synthesis
- DFT insertion
- Electric rules check (ERC) on gate-level netlist
- Floorplan
- Die size
- I/O structure
- Design partition
- Macro placement
- Power distribution structure
- Clocks distribution structure
- Preliminary check (IR drop, ESD, and so on)
- Place and route
- Parasitic extraction and reduction
- SDF generation
- Various checks
    Static timing analysis
    Cross talk analysis
    IR drop analysis
    Electron migration analysis
    Gate oxide integrity check
    ESD/latch-up check
    Efuse check
    Antenna check
- Final layout generation
- Manufacturing rule check, LVS check
- Pattern generation

35. WHAT ARE THE TWO MAJOR ASPECTS OF ASIC DESIGN FLOW?

Without question, an ASIC design flow is a complicated system that includes many commercial CAD tools, as well as many tools or scripts developed in-house. However, no matter how many tools or scripts are integrated in the flow, an ASIC design flow is yet characterized by two key purposes: create and check.

The process of creation refers to the activities of creating hardware, such as RTL coding, logic synthesis, and place and route. Figure 4.1 shows the major activities of create.

IC implementation is a process of creating hardware. The fundamental elements of an IC chip are CMOS transistors, bipolar transistors, capacitors, and resistors. In the world of ASIC integration, the basic building blocks are logic gates, memories, and special function macros. They are composed of those fundamental elements but at one abstraction level higher. The creation process has two facets: creating the chip logically and creating the chip physically. Logically, the design functions defined in the product's specifi-

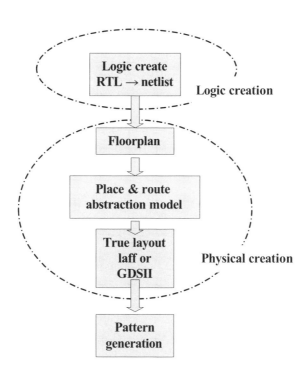

Figure 4.1. The creation process in IC implementation.

cation must be realized by real logic gates, memories, and special modules. Then this logic representation of the design is constructed physically so that a chip is produced to perform its intended functions.

The requirements for this creation process are correctness and promptness. Structure correctness must be guaranteed by the construction. This is the most important issue in implementation. This structure correctness depends heavily on the quality of the CAD tools. Promptness can be achieved by employing the hierarchical design methodology. In other words, the divide-and-conquer approach can simplify the complicated task so that the amount of design data handled at one time is significantly reduced. Promptness can also be improved by using more powerful CPUs and larger memories. Sometimes, promptness can also benefit from allocating more disk space to the project. In this way, different implementation ideas are tested in parallel, thus using less time than testing them in sequence. This approach can be characterized as "trade space with time (schedule)."

During the hardware creation process, extra effort must be taken to ensure that the resultant chip can be tested for manufacturing defects so that no bad parts are accidentally delivered to customers. This is design for testability (DFT). It is different than functional testing, which checks for functional or logic bugs that exist in all of the dies, if they exist. DFT looks for manufacturing defects that are introduced in the manufacturing process, not in the IC implementation process. The defects could exist in all of the dies and cause all of them to malfunction. But most of time, they only exist in a certain percentage of dies. In IC implementation, or the creation process, designers must pay extra attention to make their products testable.

As process geometry gets smaller and smaller, design for manufacturing (DFM) has became a subject that also needs IC implementation engineers' attention. In the creation process, further effort has to be applied to make the chip manufacture friendly so that decent yield can be reached and profit margin can be maximized.

As shown in Figure 4.1, hardware creation starts with logic synthesis, which translates the design from an RTL description to a gate netlist. Then this logic entity is transformed from a logic domain to a physical domain by physical creation. During this physical creation process, the design must go through three critical stages. In the floorplan, the chip's I/O structure, megamodule locations, power distribution structure, and the design's physical partition need to be ascertained. The step of place and route finishes the detail work of determining the cells' legal locations, routing the interconnecting wires, distributing clock signals, and so on. This step works on abstraction models of the basic building blocks (logic gates, memories, macros, and so on). In other words, during this stage of physical creation, the underlying data are not real layouts, but models.

This type of work is very algorithm-intensive and requires a huge amount of computation and a huge amount of memory storage (up to 128 G today). Using abstraction models can significantly reduce the amount of information handled by the tools and thus speed up the process and increase the manageable design size. The drawback is that the tools do not take into consideration all of the detailed information that is needed to build the chip, and sometimes mistakes/errors can be result in later stages. The last stage is the true layout creation, in which the geometries that represent the logic elements in the design are created and checked against the manufacturing rules. The IC implementation process usually ends here. The resultant layout is delivered for *pattern generation* (commonly called *PG*) or tapeout.

Checking is another significant aspect of IC implementation. The logic and physical entity created in creation process must be checked against various criteria. These include the physical check, logic check, timing check, and design integrity check. Figure 4.2 presents the various checks that a chip must pass through.

The most important check is the structural correctness check, commonly called the *layout versus schematic (LVS)* check. In an ASIC integration domain, a schematic is also called a netlist, which is the representation of the basic building elements and their interconnects. A netlist is usually derived from RTL code that captures the designers' intention. Therefore, the error-free completion of this check is the absolute necessary condition for the chip to even have a chance to work correctly. The other area of the physical check is the manufacturing rule or geometry check. This check verifies that the physical layout of the chip is constructed according to the manufacturing

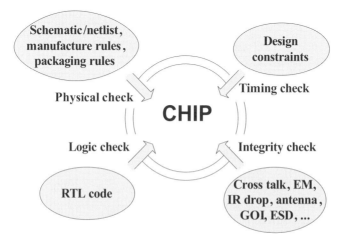

Figure 4.2. The check process in IC implementation.

rules associated with this process technology node. Complying with these rules gives the product a better chance of higher yield.

Before being sent to physical implementation, the RTL code and the resultant original netlist have already been verified for functionality. During the downstream IC implementation stages, the netlist will be modified occasionally for various reasons: clock tree insertion, buffer addition/deletion, or scan chain reordering. After each of such modifications, the new netlist must be verified against the previous netlist to ensure the logic equivalence. Preferably, this logic check should be performed after every step that changes the netlist, so that no major mistake or problem occurs in the final stage, where it is too late to be fixed. Finding bugs or problems and fixing them early is always less expensive than doing so later in the flow.

One of the widely mentioned concerns in SoC integration is timing closure. This issue is critical since the chip we build must run at the desired speed. Not only must the chip function at the desired speeds under normal working condition, but it must also do so in various other working environments, such as at different operating voltages, temperatures, transistor speeds. The timing check, or static timing analysis (STA), is the only efficient way of assuring this speed quality.

As process geometries continue to shrink, design integrity issues such as cross talk noise, electromigration, IR drop, ESD and latch-up, antenna, gate oxide integrity, and so on must be dealt with before the chip is taped out. Consequently, a design integrity check of those issues is indispensable in the chip implementation process. This check ensures that chip can be manufactured and assembled safely. It also improves the chance that the chip will work reliably for its entire lifetime.

36. WHAT ARE THE CHARACTERISTICS OF GOOD DESIGN FLOW?

Because IC implementation is closely tied to technology nodes that are constantly evolving and various new issues are emerging rapidly, a good design flow must have the flexibility to deal with those new challenges without major overhaul.

In the meantime, to overcome the number of problems presented in today's SoC implementation arena, the entire EDA industry is working at high gear. New tools are emerging at an accelerated speed; as a consequence, a good design flow should have the ability to absorb the latest developments easily.

A good design flow is flexible, not rigorous. In other words, the create and check functions in the flow should be only loosely linked. They should not hinge on each other. During the development of a complicated multimillion-gate SoC, there is much exploratory work to do before the final implementation starts: studying the die size, floorplanning, looking at run times, analyzing CPU and memory resources, allocating appropriate disk space, and so on. The purpose of these experiments is to uncover any major problems (both design- and flow-related problems) in advance. During this time, the entities created by the create function of the flow are not perfect in nature and problems are expected. Most of them can be safely discarded. Thus, the check function should not prevent the flow from executing the following operations, simply due to the fact that there are errors or problems in previous steps.

A decent design flow should also have the capability of handling large, complicated designs and small, simpler designs differently. The implementation approaches for large and small designs can be different for various reasons, such as efficiency, cost of commercial tool licenses, turnaround time, and design style (full-chip or module-level). In many situations, certain efforts or steps required for large designs are not necessary for smaller designs. Therefore, a design flow should have the option of giving control to its user or be able to configure itself on the fly for different types of designs.

In a good design flow, there should be flexibility to allow some major steps to be executed on their own or outside of the flow. Major steps such as place and route, static timing analysis, and logic equivalence check require detailed tunings for different designs. Sometimes, it is more efficient to run these operations outside of the flow.

A good design flow should be friendly to *engineering change order (ECO)*. ECO, which is inevitable in large designs, should require only a minimum effort from the design flow point of view.

Finally, a good design flow should treat safety as its top priority. For a design flow that is used widely in an organization, hundreds of projects could be executed each year through the flow. Therefore, the flow must guarantee that the projects passed through it have a high possibility of success. In other words, the check function of the design flow must be robust, sometimes even at the expense of efficiency.

37. WHAT IS THE ROLE OF MARKET RESEARCH IN AN ASIC PROJECT?

A project should never be started without extensive market research. Without an understanding of market needs and technical feasibility, the project is virtually guaranteed to fail. Good market research should result in a clear product definition for the targeted market. There are many examples of technically excellent products that end up as financial disasters simply because they are either too early or too late to market, or are designed for markets that cannot support the volume of product required to provide a capital return. Market research should direct the product definition to the market needs, at the appropriate time. This research should also verify the value and the size of the market opportunity.

Good market research is vital for technical and financial success.

Many ASICs are customer specific. In such cases, it is imperative to discuss the needs directly with the customer. By communicating with the customer directly, the design and management team can more readily grasp the project complexity and market requirements. A secondary approach to be taken in addition to customer discussions is to study competitors' products, datasheets, and other sources of data such as those obtained at trade shows.

It is extremely important that the status of the market and the project be monitored continually. If market conditions change or an unforeseen technological difficulty is encountered, it must remain a possibility that the project can be terminated. Perhaps the toughest of all decisions that have to be made during the execution of a project is whether to kill the project if conditions change. Despite the loss of a great deal of work, this must be done dispassionately. This is where the expertise of sound, experienced, and unemotional managerial decision making is important. Poor management can cause more severe losses than would otherwise occur, by allowing continuation of projects that should be killed and potentially causing friction with customers.

In the IC design business, the market governs everything!

38. WHAT IS THE OPTIMAL SOLUTION OF AN ASIC PROJECT?

The optimal solution of an ASIC project is producing a chip that fulfills its full functional requirement using the least amount of resources possible. In detail, this definition includes the following goals:

- The process technology chosen is just right for the project. It is not too advanced with difficult integration challenges or a high price tag, nor is it obsolete with speed degradation or a large silicon area penalty.
- The functional blocks included in the system (chip) are just adequate for the targeted market. The chip defined should have all of the func-

tions needed to attract as many potential customers as possible. On the other hand, no unused functions should exist on the chip that waste resources.

- The silicon area used by the final implementation is precisely what is needed for the desired operating speed, the design-for-testability requirement, the appropriate power and clock distribution network, a respectable chip reliability margin, and the design-for-manufacturability requirement. In other words, there should be no empty or unused silicon area on the chip.

- The chip is 100% testable for mass production.

- The clock and power distribution networks are neither overdesigned, with large, unnecessary safety margins nor underdesigned, with the risk of malfunction or reliability weakness.

- The chip is designed in such a way that while power consumption is kept at a minimum for all of the benefits of low-power operation, it is not done so at the expense of its normal functionality and operating speed.

- The package is designed just right for the chip's pin-count and power consumption. The cost should be kept to a minimum.

To reach this optimal solution for a given ASIC project requires a huge amount of execution effort, which involves engineering resources, hardware equipment, tool license investments, and execution time. As depicted in Figure 4.3, the ideal, or optimal, solution is imaginarily represented by a horizontal line. The dots represent individual design implementations (solutions). The distance between any dot and the optimal solution line represents the quality of that implementation, measured by the combined qualities of each submeasurement against the above-mentioned criteria: the greater the distance, the poorer the implementation (solution).

It is usually true that during the first few tries, the quality of the implementations is far from optimal. Those that follow have a better chance of approaching the ideal because information about the design accumulates from previous mistakes and errors. However, to reach the absolute optimal solution requires a tremendous amount of execution effort, which is usually beyond reasonableness. For real design projects, especially for products targeting the consumer electronic market, if the project execution time passes a certain time milestone the product will completely miss the market window and loss its validity for continuation. Therefore, unlike science or research projects, it is often unwise to search endlessly for the optimal design implementation. In most cases, a "good" technical implementation, not necessari-

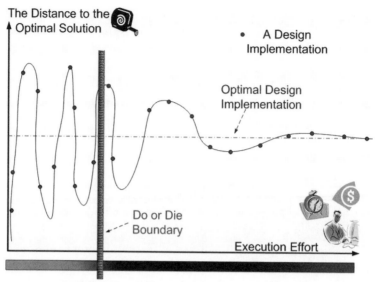

Figure 4.3. The optimal solution and execution effort.

ly the optimal one, can turn out to be the best solution if judged according to overall project profitability.

39. WHAT IS SYSTEM-LEVEL STUDY OF A PROJECT?

After a product is successfully defined from marketing research, it is often followed by a *system-level study*. Sometimes this system-level study can start early during the production definition phase.

The main focus of a system-level study is feasibility analysis. For a typical SoC application, this study addresses issues in three critical areas: algorithm, architecture, and software integration. Algorithm design is an essential task in signal processing applications such as wireless telephony, multimedia codecs, DSL, and cable modems. The optimal algorithms should meet the design functional requirements with minimum resource claim. Architecture design is the work of putting together the right processors, custom logic, on-chip buses, memories, and peripherals in order to make most effective use of the silicon. It is performed through abstract modeling of the SoC architecture, which consists of those processors, algorithms, custom logic, buses, memories, and peripherals. The goal is to find the optimal architecture based on the trade-off between software tasks and hardware functions with respect to performance, throughput, and latency. Software integration addresses the analysis of the interaction between the hardware and software components of the SoC.

System-level design does not involve implementation detail. It is the approach of viewing the chip in a big-picture perspective. It abstracts away the full detail of the design, retaining just enough features to validate that functions embodied by the design can perform the specified design goal and can satisfy the performance criteria. The aim of this high-level study is to ensure that the chip is built on a solid foundation, the chip is constructed fundamentally correct, and the chip architecture is the optimal solution based on the trade-off between performance and resources required. Preferably, the system-level study and modeling should also support the smooth migration to downstream implementation.

40. WHAT ARE THE APPROACHES FOR VERIFYING DESIGN AT THE SYSTEM LEVEL?

The approaches used in system-level study can be roughly classified as: algorithmic, modular, cycle-accurate, and RTL.

In an algorithm-level study, only the behavior of the design, not the specific implementation detail, is specified, whereas in a modular-level study, the design is partitioned into components that communicate through clearly specified protocols. A cycle-accurate level study introduces the notion of the clock and the times at which events occur, but it does not completely specify the implementation details of the events. An RTL-level study specifies the implementation of the events but without relying on any particular implementation technology.

System design engineers usually first describe and simulate their systems in the C/C++ language to study system behavior at the algorithm level. Then they will move one step further toward implementation by describing and simulating the system in a *hardware description language (HDL)* such as Verilog or VHDL. However, as design size becomes larger and larger and design complexity reaches higher and higher levels, a new trend of using *electronic system language (ESL)* to replace C/C++ and RTL as the system tool is slowly emerging. The two most promising languages in this field are systemC and systemVerilog.

ESL is a new approach in the IC design regime. The transition from the RTL-level to ESL will not be abrupt. It will occur more as an evolution than a revolution. This transition will be along the lines that software industry followed as it evolved from the strict use of hand-coded assemblers in the

1950s to the extensive use of compiler in the 60s of last century. At first, only the noncritical portions of time-to-market design will be affected by ESL. Over time, more sophisticated compilers and synthesis algorithms augmented by increasing hardware functionality will extend the reach of these ESL automatic techniques until only extremely performance-driven designs must be implemented at the RTL level.

The benefits of system-level study include the following:

- A higher level of abstraction reduces design time.
- A higher level of abstraction means faster verification.
- Abstraction and encapsulation lead to reuse, creating more gates in less time.
- A common language for hardware and software promotes hardware–software codesign.

41. WHAT IS REGISTER-TRANSFER-LEVEL (RTL) SYSTEM-LEVEL DESCRIPTION?

As addressed in Question 40, after we study the system at the algorithm, modular, and cycle-accurate levels, we need to move forward in the direction of hardware implementation. Describing the system at the register-transfer-level (RTL) will bring us one step closer to this goal.

RTL, or *register transfer level,* is a method of describing hardware behavior in software using a hardware description language (HDL). In any hardware description language (Verilog or VHDL), there are two different approaches to describing a hardware block. One method is to describe it by its behavior only, with no consideration of how to achieve the intended behavior in hardware. The other approach is to describe the system in a structural way, or to achieve the intended functions with basic building blocks whose functionalities are already known and well defined. In this approach, a Verilog or VHDL description of the system is created at a time when hardware engineers know roughly what gates they want and where they want them. This is only possible after the cycle-accurate level study is finished and the block's clocking structure is finalized. This structural RTL study, also called synthesizable RTL, does not have to tie to a particular technology library; it only needs to reach the level of generic logic functions and storage elements (register, latch, memory).

Describing the system at the RTL is a powerful tool in designing an IC chip. It is also an absolutely necessary step for ensuring system correctness since the RTL description can be simulated intensively by many simulators.

However, RTL is mainly used for capturing digital design and developing digital IP. It is not a tool targeted for transistor-level design (analog, mixed signal design). Compared to transistor-level or gate-level simulations, RTL's higher abstraction level can produce faster simulations, which is crucial in large-scale digital design.

42. WHAT ARE METHODS OF VERIFYING DESIGN AT THE REGISTER-TRANSFER-LEVEL?

There are several approaches for verifying the functional correctness of a RTL-described chip design:

- Software-based simulation
- Hardware-based simulation acceleration
- FPGA hardware prototyping
- Formal verification

Simulation is the most commonly used method of verifying RTL code. In addition to performing simulation, most RTL simulators have the additional capabilities of lint checking, state machine analysis, assertion checking, and code coverage analysis. These tools can help RTL designers achieve cleaner code and better test benches. The major shortcoming of simulation approach is that it is not a thorough verification. For most of designs, it is impossible to cover all of the application scenarios.

Simulation speed is a big concern in large designs. Therefore, there are certain types of hardware dedicated to the purpose of simulation acceleration. Those emulation boxes can read in the RTL code and simulate it at much higher speeds.

Another approach is to synthesize the RTL code into field programmable gate array (FPGA) hardware and prototype the design. This alternative can give the verification task a significant speed boost. The FPGA prototype can be plugged into the end system for testing in real applications. The circuit speed of this kind of FPGA prototype is often much slower than that of the final ASIC, but it can improve the designers' confidence in the functional correctness of the RTL code.

Formal property verification is enjoying growing importance as digital designs get more complex and traditional validation techniques struggle to keep pace. Formal verification methods include *symbolic model checking (SMC)* and *theorem proving (TP)*. In symbolic model checking, the temporal logic specification is used to check system properties; the system is modeled as a finite-state machine. For theorem proving, both the system and its desired properties are expressed as formulas in certain mathematical logic and the theorem prover will find a proof based on the axioms defined for the system. Formal methods introduce mathematical rigor in their analysis of digital designs. Consequently, they can guarantee exhaustive coverage. Within this framework, compared to the simulation approach, designers can employ solid design abstraction techniques to manage the complexity.

43. WHAT IS TEST BENCH?

Functional verification is one of the bottlenecks in delivering today's highly integrated IC chips. Verification complexity tends to increase exponentially with design size. The continued advancement in the level of SoC integration has placed an enormous burden on today's verification engineers. Despite the significantly harder verification task, they must continue to ensure that no bug is missed when the design is delivered for manufacturing.

In an IC design regime, the term *bug* is used to describe a design error that is introduced during the design process unintentionally. A bug often results in unexpected system behavior, which causes functional error, or worst of all, chip failure. The task of verification is to find all of the bugs in the system. The tool of choice in this task is the test bench created for simulating the design.

A test bench is an entity constructed in HDL, such as VHDL and Verilog, or in some other higher-level languages. It stimulates the module (device) under test and observes its behavior. As shown in Figure 4.4, the test bench is the driver that provides the stimulus to activate the device. It also captures

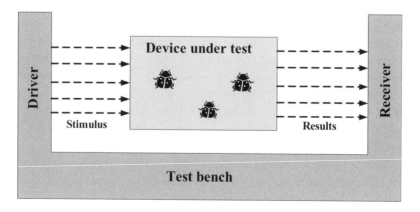

Figure 4.4. Test bench.

the behavior of this device under this set of stimuli to evaluate its performance. If the test result does not agree with what is expected, then there are potential functional errors, or bugs.

The test bench has become an integral part of the IC design process. Its aim is to ensure that the HDL module is sufficiently tested, or no known bug exists, before it can be implemented in hardware. The most challenging part of the test bench creation process is to produce a set of test benches that can cover all the application scenarios, or as many as possible.

44. WHAT IS CODE COVERAGE?

Code coverage is closely tied to the concept of the test bench. It is a measurement of the quality of the test bench. Using a particular test bench, the code coverage of a module constructed in HDL (or other higher-level languages) records which lines of the RTL source code are executed and which are not. The premise is that it is impossible to catch any bugs lurking in it if a line has never been executed by the test bench.

Code coverage analysis is a structural testing technique that compares test bench behavior against the apparent intention of the source code. It assures the quality of the test bench, not the quality of the actual source code or the actual module. Code coverage analysis is the process of finding areas of a source code not exercised by a set of test cases in the test bench. It can help to create additional test cases to increase the coverage. This analysis gives a quantitative measure of the coverage, which is an indirect measure of the quality. It can also identify redundant test cases in the test bench that do not increase coverage. Ultimately, the result of code coverage analysis of a particular design will impact a designer's level of confidence in his or her RTL code.

45. WHAT IS FUNCTIONAL COVERAGE?

As addressed in Question 44, code coverage analysis is a structural testing technique. *Functional coverage,* which contrasts to code coverage, compares test bench behavior to a product specification. Structural testing examines how the source code works, taking into account possible pitfalls in the structure and logic. Functional testing examines what the source code intends to accomplish, without regard to how it works internally. Code coverage analysis could be viewed as glass-box or white-box testing, whereas functional coverage analysis could be regarded as black-box testing.

The objective of functional coverage is to maximize the probability of simulating and detecting functional faults, or bugs. It attempts to achieve this at minimum cost in terms of time, labor, and computation. It is difficult to formally create a functional coverage metric and prove that it provides a good proxy for finding bugs. However, the very fact that functional coverage metrics can improve the chance of detecting bugs seems empirically true based on observations of and experiments involving many SoC projects.

46. WHAT IS BUG RATE CONVERGENCE?

In the verification process of an IC design project, functional testing (by using a test bench) helps locate design errors, or bugs. This verification process has three phases: constructing and bringing up the test bench, verifying the basic test cases, and verifying the corner cases.

During the first two stages, bugs are easily detected and, consequently, the rate at which bugs are found is relatively high. Then, as the design gradually becomes mature and verification continues into corner-cases testing, bugs become harder and harder to find, and this rate correspondingly slows down, as shown in Figure 4.5. Eventually, when verification is almost complete, the bug rate is virtually zero. The metrics of *bug detection frequency, length of simulation after last bug found,* and the *total number of simulation*

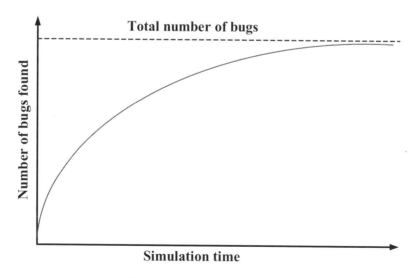

Figure 4.5. Bug rate convergence.

cycles are the most commonly used methods to measure the confidence level of the overall verification process.

47. WHAT IS DESIGN PLANNING?

Design planning is the process of defining the chip's implementation characteristics. It is a task that is carried out before the real implementation work begins. It includes the following major tasks.

- Selecting the technology and library. Choosing the appropriate technology node based on process speed, price, IP availability, foundry availability, and so on. For the chosen technology, selecting the appropriate ASIC library based on design requirements, such as high speed or low power.
- Selecting the IP. Choosing the style of IP: hard or soft; the source of IP: internal or external. Studying the availability of on-chip memories. Studying the availability of various special functions on the chosen technology, such as PLL, DAC, or ADC.
- Selecting the I/O based on speed, drivability, power usage, voltage interface, and special I/O functions such as DDR, SERDES, and USB.
- Selecting the package. Consider the number of available pins, the price, the footprint, the maximum power rating, and the thermal characteristics.
- Estimating die size. Conducting preliminary floorplanning and determining whether the design is I/O limited, core limited, megamodule limited, or package limited.
- Partitioning the design. Determining whether the design should be implemented in top-down or bottom-up fashion. And, for large designs with high gate counts, the design must be divided into subchips or hierarchical modules and implemented in the manner of divide and conquer.

A hierarchical approach can enable faster implementation, but often at the price of nonoptimum result as compared to a flat approach. All of the tools used nowadays in the implementation process, especially the synthesis tool, have component limits, due to the constraints of CPU processing power and memory capacity. With the capability of today's EDA tools, it is a good idea to limit the number of components in an ASIC design netlist to less than one million.

During the design-planning phase, these issues are critical design decisions. Together, they lay down the foundation for the following physical implementation work. Any serious misjudgment will result in a severe consequences, such as a complete reset of the implementation cycle.

48. WHAT ARE HARD MACRO AND SOFT MACRO?

Hard macro and *soft macro* approaches are the two methodologies by which IPs are delivered. The hard macro method is used to transfer an IP block that has not only the logic implementation but also the physical implementation. In other words, the physical layout of a hard macro IP is finished and fixed in a particular process technology. Meanwhile, a soft macro IP has only the logic implementation without the layout.

The biggest advantage of the hard macro approach is optimization. The hard macro block is timing-guaranteed and layout-optimized. The drawback is poor portability since it is already tied to a specific process technology. In contrast, a soft macro IP has excellent portability. It can be synthesized into any ASIC library if the RTL code and design constraint are available.

Typically, IPs with significant analog content are delivered as hard macros since analog IPs are very process and layout sensitive. Digital IPs, on the other hand, can have the flexibility of being "hard" or "soft." Furthermore, for digital IPs, there is another type referred to as netlist IP. Netlist IP has only the gate-level netlist but not the RTL code, mainly for security reasons. The netlist IP can be used for the same process or library, or it can be used for porting and thus mapped into a different process or library. Obviously, the drawback of the soft macro approach is the extra work of physical implementation, or layout. Compared to hard macro, the verification of soft macro also requires more attention.

49. WHAT IS HARDWARE DESCRIPTION LANGUAGE (HDL)?

The design of integrated circuits (ICs) is an art. During a half century of IC development, it has gradually become clear that there is a need for a computer language to describe the structure and function of integrated circuits, or for describing an entire electronic system. In the 1980s, the need for such a *hardware description language (HDL)* finally drew the attention of the government, the electronics industry, and universities. As a result, two HDL languages, Verilog and VHDL, have been standardized and have emerged as the tools for IC design.

An HDL is created to meet a number of needs in the IC design process. First of all, it allows the description of the structure of a hardware system. An HDL can be used to describe how the system is decomposed into building blocks and how those building blocks are interconnected. Second, it allows the specification of the system functionality by using the form of familiar programming language. Third, the design of a system can be simulated before being manufactured so that designers can quickly compare alternatives and test for correctness without the delay and expense of hardware prototyping. Fourth, it allows the detailed structure of the design to be synthesized from a more abstract specification, allowing the designers to concentrate more on strategic design decisions. This automatic synthesis process also helps reduce design implementation time.

Overall, the use of HDL can benefit the IC design process in following aspects: documenting the design, simulating the behavior of the design, and directly synthesizing the design into real hardware.

50. WHAT IS REGISTER-TRANSFER-LEVEL (RTL) DESCRIPTION OF HARDWARE?

Using an HDL to describe a hardware system is carried out on three levels: the gate level, the register-transfer level (RTL), and the behavioral level.

The gate level describes a system in a purely structural fashion, by decomposing the system into basic building blocks whose functionalities are well defined and whose structures are fixed. It is not easy to extract the functional sense by just reading gate-level HDL since it only contains components and their interconnections. Gate-level description is primarily used in the last stage of IC implementation to precisely describe the physical structure of the design, with little or no attention paid to its functionality. On the other hand, behavioral-level description does not pay any attention to the implementation aspects of the design. It simply describes the behavior, or functionality, of the design by using a higher level of abstraction, with no information or direction on how the design will be implemented.

The RTL level of description is somewhere in between. It defines the system behavior by describing how the data, or information, is transferred and manipulated inside the system. It implies the system structure by referring directly to the data storage elements and describing how the data should be manipulated between those storage elements. However, RTL code does not go further than this to directly specify implementation details, such as which sequential cell to use for data storage or what logic cells to use for data manipulation.

From RTL code, an experienced designer can extract functionality; a synthesis tool can implement it into a physical entity (netlist) when an ASIC library is selected. Compared to gate-level description, RTL-level description describes a design at a higher level of abstraction. It encourages the designer to focus on the functionality of the design rather than on its implementation, while leaving the automatic synthesis tool to realize and optimize the functionality specified. In other words, RTL allows the designer to describe "what" the design does, and lets the synthesis tool decide "how" the design should be implemented in order to create the optimal implementation. As a matter of fact, the specific HDL coding style required by Synopsys' synthesis tool (e.g., Design Compiler) is referred to as RTL coding.

Figure 4.6 is an example of RTL code, composed in VHDL language. This section of RTL code describes a hardware block of a frequency divider. The divide ratio is 10. Figure 4.7 is the symbol of this block, which shows an input pin *INCLK* and an output pin *OUTCLK*. If a signal of frequency f is presented at input pin *INCLK,* a signal bearing frequency $f/10$ is generated at output pin *OUTCLK*. This behavior is verified by simulation, as shown in Figure 4.8. There is one cycle of output clock for every ten cycles of input clock.

51. WHAT IS A STANDARD CELL? WHAT ARE THE DIFFERENCES AMONG STANDARD CELL, GATE-ARRAY, AND SEA-OF-GATE APPROACHES?

Standard cells are the basic building blocks of cell-based IC design methodology. A standard-cell library is one of the foundations upon which the ASIC design approach is built. A standard cell, from a library, is designed either to store information or perform a specific logic function (such as inverting, a logic AND, or a logic OR). The type of standard cell created to store data is referred to as a sequential cell. *Flip-flops (FF)* and latches are examples of sequential cells, which are indispensable elements of any ASIC library. The type of standard cell used to perform logic operations on the signals presented on its inputs is called combinational cell.

Standard cells are built on transistors. They are one abstraction level higher than transistors. As shown in Figure 4.9, a hardware block can be represented in four different abstraction levels during the chip implementation process. The lowest level is the transistor or device level. At this level, the entire block is described directly by the very basic building elements of transistors, diodes, capacitors, and resistors. One level up is the cell level, in which designs are composed of standard cells. One more step up is the mod-

```
LIBRARY IEEE;
use IEEE.std_logic_1164.all;
use IEEE.Std_Logic_unsigned.all;

entity DIV10 is
          port (INCLK : in std_logic;
           OUTCLK : out std_logic
          );
end DIV10;

architecture RTL of DIV10 is
signal counter: std_logic_vector(3 downto 0) := "1011" ;
begin
process
begin
 wait until INCLK'event and INCLK = '1';
          if ( counter = "1001" ) then
                     counter <= "0000" ;
          else
                     counter <= counter + '1' ;
          end if;
 end process;

process
begin
 wait until INCLK'event and INCLK = '1';
          if ( counter <= "0100" ) then
                     OUTCLK <= '1' ;
          else
                     OUTCLK <= '0' ;
          end if;
 end process;

end RTL;
```

Figure 4.6. The RTL description of a hardware block (frequency divider).

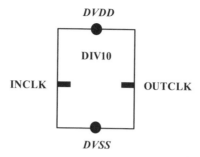

Figure 4.7. The hardware block symbol.

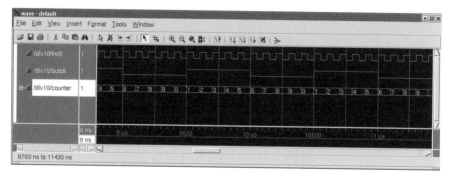

Figure 4.8. The simulation of the hardware block.

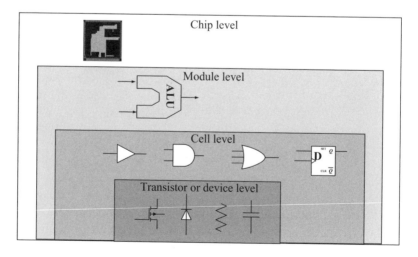

Figure 4.9. Level of abstraction.

ule level. At this level, designs are represented by modules such as adder, multiplier, ALU, and shifter. The highest level is the chip level. At the chip level, designs are partitioned into subsystems, such as DSP, microcontroller, MPEG decoder, UART, USB, DMA, ADC, DAC, and PLL. The higher the abstraction level, the less implementation detail it contains. Levels of standard cells are created for easy chip implementation, especially for large digital designs.

During the chip construction process, a designer's HDL code is transformed to a netlist using synthesis tools. The resultant netlist is composed of a certain number of standard cells, each one having its specific logic function. Overall, the intended system functions, initially described by HDL code, are realized by the combined logical effect of all the standard cells in this netlist. Down the road of implementation, these standard cells are placed within the chip's floorplan by special place and route tool. The interconnections among these standard cells are also routed and wired by this tool.

Figure 4.10 shows several standard cell examples. The top row has three combinational cells of an inverter, a two-input OR gate, and a two-input NAND gate. The bottom row is one sequential cell of D-type flip-flop. There are many associated views of any standard cell in a library, for various purposes. These different cell views are used by different EDA tools for schematic capture, simulation, timing verification, place and route, power analysis, and electrical check during the chip implementation process. In this figure, the three most common cell views are shown: symbol view, schematic view, and layout view. From this schematic view, it is clear that different cells have different structure complexity depending on their logic functions. The simplest cell (the inverter) contains only two transistors. On the other hand, the D flip-flop has 26 transistors embedded in it as its functionality is much more complicated than that of an inverter. Correspondingly, the size of its layout is much larger than that of an inverter as well.

Figure 4.11 shows more details of the OR gate layout. The cell's physical size is defined by cell height and cell width. The cell boundary is an attribute used by a place and route tool to place the cells during the placement stage. At the top and the bottom of the cell, there are usually wide strips of metal for the power supply and ground connections (*DVDD* and *DVSS* ports, as shown). The transistors inside the cell are formed by the geometries on the base and metal layers as represented by the different shadings in the figure (see also Chapter 2, Question 14). The exact shapes and dimensions of those geometries are drawn in the layout view. Typically, only the metal1 layer is used inside the cell layout since higher level metals are reserved for signal routing. In the layout view, there are also geometries that define the cell's signal ports (A, B, and Y for this OR gate) that will be used

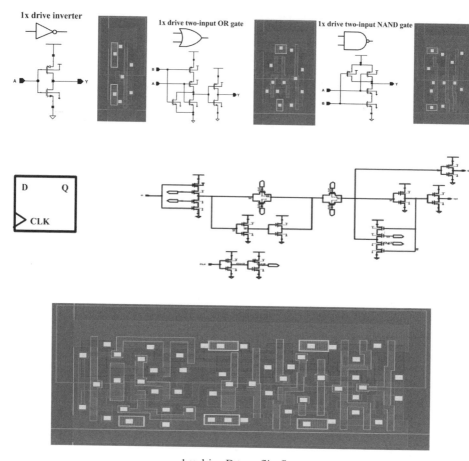

1 × drive D type flip-flop

Figure 4.10. Several standard cell examples.

later by the router to make the signal connections. It is critical to complete each standard cell's layout with the least amount of silicon area possible. The reason is that the number of standard cells in a design could be on the order of millions, and a small amount of overuse of area at the cell level could add up to a large penalty in area at the chip level. In some cases, the same logic cell could have different layout versions: for example, small layout footprints for low-performance but compact designs and large layout footprints for high-performance and expensive projects.

Physically, the standard cells within an ASIC library have a fixed size in one dimension (usually the height) so that they can be placed and aligned along the rows of the chip. Figure 4.12 shows two rows of standard cells

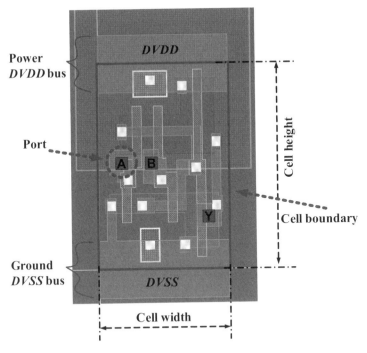

Figure 4.11. The cell layout view in detail.

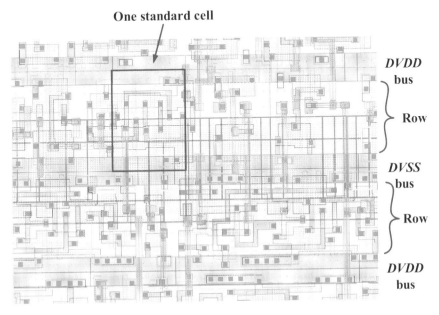

Figure 4.12. Two rows of standard cells in a chip layout.

within a certain portion of a chip layout. Each of those rows is filled with various standard cells that make up the actual chip. In some special cases, certain standard cells are designed twice (or even more) as high as regular cells. But those heights must be multiples of the regular cells so that they can be placed appropriately in the rows by the place and route tool. As seen in Figure 4.12, when the cells are placed next to each other in a row, the *DVDD* and *DVSS* geometries abut to form one long metal strip (see also Question 72). Also, between two adjacent rows, the cells are flipped vertically so that *DVDD* and *DVSS* metals are shared among the two rows. As a result, there is no wasted space existed between the rows. A large chip has a huge number of rows with power and ground busses running through each row. In this configuration, the height of the row (which is also the height of regular cells) is often referred to as row pitch and the smaller the pitch, the higher the gate density. As process technology continuously shrinks, so does the size of the standard cells. As a result, more and more logic gates can be packed into the same silicon area.

The *gate-array* approach is another implementation methodology for digital design. A gate-array circuit is a prefabricated silicon chip with no particular function predefined. In this prefabricated silicon, transistors and other active devices are placed at regular, predefined positions and manufactured on a wafer, called a master slice. Creating a logic gate with a specified function is accomplished by adding metal interconnect layers to the chips on the master slice late in the manufacturing process. These logic elements are then joined by the metal layers to customize the function of the chip as desired. Typically, the base layers (well, diffusion, and polysilicon) are fixed and prefabricated. Metal layers are added in later stages to configure (personalize) the chip to specific functions based on user requirement. Figure 4.13 shows three gate-array cells, two of which are unprogrammed and one is already personalized by extra metal geometries (a three-input NAND gate). These gate-array cells may be arrayed in rows as shown at the bottom. There are routing channels reserved between the rows for cell-to-cell signal routing tracks.

Figure 4.14 is the *sea-of-gate* design style, which is a generalization of the gate-array approach. In this style, continuous rows of n and p diffusion are run across the master chip. These in turn are arrayed regularly in a vertical direction without routing channels in between. A logic gate is isolated from neighboring gates by tying the gate terminal of the end transistors to V_{SS} (NMOS) or V_{DD} (PMOS). Cell-to-cell signal routing is achieved by routing across rows of unused transistors. This can produce a much denser general-purpose array. In this figure, an isolated three-input NAND gate is programmed.

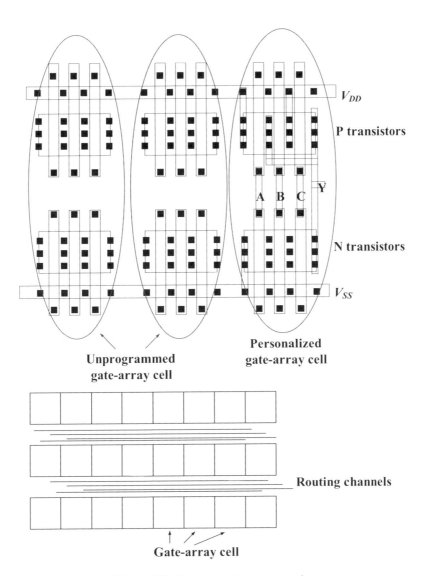

Figure 4.13. Gate-array design approach.

The greatest difference between the standard-cell approach and the gate-array and sea-of-gate approaches is that the masks for all the layers are required to construct the chip using the standard-cell approach, whereas in the gate-array or sea-of-gate approaches only the metal layers are required (the base layers are prefabricated). Gate-array and sea-of-gate master slices are often prefabricated and stockpiled in large quantities regardless of customer orders. As a result, the design and fabrication according to the individual

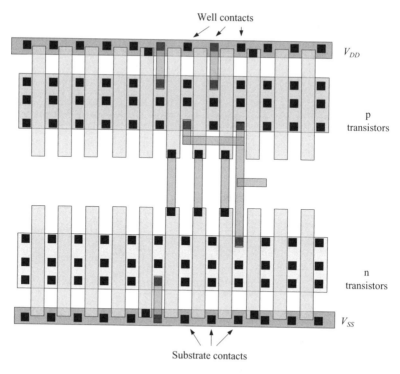

Figure 4.14. Sea-of-gate design style.

customer specification may be finished in a shorter time compared to standard-cell design. The gate-array approach can reduce mask costs because fewer custom masks are required. Additionally, costs and the lead time for manufacturing test tooling are reduced as well since the same test fixtures are used for all the gate-array products manufactured on the same die size. The drawbacks of the gate-array and sea-of-gate approaches are low density and inferior performance compared to standard-cell design. However, they are often a viable alternative for low-volume products.

For very high performance digital designs and most analog designs, the full-custom style is preferred. Figure 4.15 shows one example of a full-custom layout in which every single transistor is designed and laid out manually.

In summary, the level of standard cells (gate-array and sea-of-gate) is created so that the task of chip implementation, especially for large SoC designs, can be carried out efficiently. Although it is very important, the standard cell is not one of the basic building elements of the silicon chip. The basic building elements of a CMOS chip are NMOS and PMOS transistors, diodes, capacitors, resistors, and inductors. During the very last stage of the

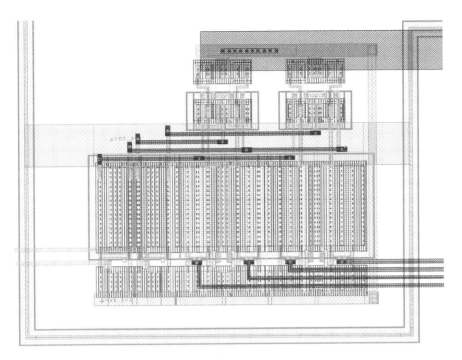

Figure 4.15. Full-custom layout.

chip implementation process, the standard cells are decomposed into NMOS and PMOS transistors, and the standard cell boundary then disappears. However, without standard cells, every single chip design must start at the transistor level. This degrades the efficiency of implementation greatly or it could make the task impossible.

52. WHAT IS AN ASIC LIBRARY?

An *ASIC library* is a group of standard cells glued together as a package. Typically, an ASIC library contains a sufficient number of combinational cells to perform any logic operation required by commonly used design styles with decent efficiency. It should also have many types of sequential cells to meet any storage requirements.

A typical modern ASIC library could have more than several hundred different standard cells. Those cells are categorized into groups by their functionality, such as INV, BUF, NAND, NOR, AND, OR, XOR, Boolean functions, flip-flop, and scan flip-flop. Within each functional group, there are a certain number of cells with different drivability. For example, in the inverter group of INV, there are usually a 1× drive cell INV1; a 3× drive cell INV3; and INV5, INV7, and INV9 cells. It is also possible to have half-drive cells in a library targeted for high density. Different drivability cells within one group provide flexibility to the synthesis tool so that the optimal result can be achieved. Nowadays, there is also a trend to create the cells within the same function group with linearly increased drivability. Such a linear ASIC library can be better utilized by the "logic effort"-driven design implementation methodology.

An AISC library is often tied to a particular process technology. Sometimes, several ASIC libraries can coexist within the same process technology, each targeting a specific purpose such as high speed, low power, or high density.

Because standard cells are the basic building blocks of ASIC design methodology, appropriate information must be provided to the CAD tools when they are used to create a silicon chip: the cell's physical appearance, its logic functionality, timing behavior, and electrical characteristics. For this reason, the cells in the ASIC library are characterized, modeled, and packed in certain data formats. A complete ASIC library should have following information (which is characterized as a view) available for each cell for the automatic design tools to use during various design phases: logic view, timing view, physical view, power view, and electrical view. Together, these views provide a complete picture of each cell in the library. Various automatic CAD tools used in IC design implementation utilize them to make their design decisions.

The quality of an ASIC library has a great impact on the quality of the designs that use this library. There are several criteria upon which ASIC libraries are studied and judged:

- Efficiency. A library is efficient and high quality if the resultant synthesized and placed-and-routed designs are fast and small, and consume less power.
- Robustness. A library that yields a good balance of area, power, and performance is of no use if it is not reliable. Issues related to ESD tolerance, latch-up prevention, electromigration, GOI, antenna, and noise sensitivity must be considered.
- Portability. Another aspect of increasing importance in evaluating a library is its adaptability for multiple foundries. A library with good adaptability is built around a common set of rules that accommodates the design rules of several silicon vendors.
- Usability. A great library is useless without design kits. Depending on the design methodology, the following types of design kits may be re-

quired: schematic capture, synthesis, simulation, place and route, static timing verification, and ATPG.

- Timeliness. The library must be available early in the technology life cycle. Otherwise, this technology cannot be used to its full potential to generate revenue.
- Cost. The cost of a library is tricky. The cost of developing a library internally can be high, both in terms of development tools and time. However, buying an off-the-shelf library always includes a compromise between what is desired and what is available. Therefore, between buying and developing, there must be a clear economic advantage for one over another, or vice versa.

53. WHAT IS LOGIC SYNTHESIS?

Logic synthesis is the process of translating an abstract form of desired circuit behavior (typically in RTL) into a design implementation in terms of logic gates (standard cells). This process is carried out by automatic synthesis tools with sophisticated algorithms. The outcome of this logic synthesis is the netlist, which is composed of various standard cells and special macro cells. The functionality of this netlist should agree with what is described in the original RTL code. Logic synthesis is one major aspect of electronic design automation.

During the process of logic synthesis, starting from the RTL description of a design, the synthesis tool first constructs a corresponding multilevel Boolean network. Next, this network is optimized by using several technology-independent techniques. The typical cost function used during technology-independent optimizations is the total literal count of the factored representation of the logic function, which correlates well with real circuit area. Finally, technology-dependent optimization transforms the technology-independent circuit into a network of gates in a given technology (library). The simple cost estimation performed in previous steps is replaced by a more concrete, implementation-driven estimation during and after technology mapping. Mapping is constrained by several factors, such as the availability of gates (logic functions) in the technology library; the drivability of each gate in its logic family; and the delay, power, and area of each gate.

In addition, behavioral synthesis is a method of synthesizing logic from a circuit specified at the behavioral level by an HDL. This approach transforms a behavioral HDL specification into an RTL specification that is then used for gate-level logic synthesis. The goal of behavioral synthesis is to increase the designer's productivity to meet the challenge of ever-increasing design size.

It is important that the logic synthesis process be:

- Free of misinterpretation
- Fast in execution
- Capable of handling large designs

Further, for the resultant netlist to deliver high-quality circuits, the area must be small, the power consumption needs to be low, and circuit speed should be high.

The quality of a logic synthesis task is highly depended upon the ASIC library used, the algorithms embedded in synthesis tools and the CPU, and the memory configuration of the computer that carries out the synthesis task.

54. WHAT ARE THE OPTIMIZATION TARGETS OF LOGIC SYNTHESIS?

The optimization targets of a logic synthesis task are speed, area, and power. For successful implementation into real circuits or logic gates, a circuit produced by HDL synthesis must meet certain speed requirements defined by clock frequency. Otherwise, it is less effective, if not totally useless. Another obvious goal of logic synthesis is that the resultant circuit should occupy as little silicon area as possible to maximize profit margins.

The third concern in logic synthesis is power consumption. Nowadays, as more and more IC designs are targeted for mobile applications, a chip's power usage has become a very sensitive issue. However, as process geometries keep getting smaller, the leakage current problem becomes more severe and starts to gradually move from backstage to center stage. In other words, lower and lower power consumption is desired on one hand, yet power managment becomes tougher and tougher on the other hand. As a result, optimizing for power is a more serious and challenging logic synthesis issue than ever before.

These three optimization targets are not isolated, but related, to each other. During the optimization process, trade-offs are often made among these targets. For example, to achieve higher circuit speed, more silicon area might be required and more power needed. To achieve better power control, more gating circuitry might be required, which, in turn, requires more area.

55. WHAT IS A SCHEMATIC OR NETLIST?

A *schematic,* or *netlist,* is the real circuit representation of an electronic design. It consists of the basic circuit elements (*instances*), their interconnections (*nets*), and certain attributes. Compared to a RTL description or design specification, which only address the designer's design intention, a schematic or netlist is the circuit representation that is one step closer to real implementation in silicon.

Structurally, a netlist either contains or references the descriptions of the components (such as logic gates, special macros, etc.) or devices (such as transistors, capacitors, resistors, etc.) it uses. Each time a component or device appears in a netlist, it is called an instance. Each instance has a master which is the definition that lists some of its basic properties and the connections that may be made to it. These connection points are called *ports* or *pins.*

In large designs, it is common practice to split the design into pieces. Each piece then becomes a definition that can be used as an instance in a higher level of the design. This approach of netlisting is called *hierarchical description.* A definition that includes no instances is referred to as a *primitive* or *leaf cell,* and a definition that includes instances is hierarchical. *Flat* designs are those in which only instances of primitives are allowed. Hierarchical designs can be exploded or flattened into flat designs via recursive algorithms. The advantage of the hierarchical approach is that it reduces the redundancy in a netlist, saves disk or memory space, and produces a design that can be more easily read by both machines and people.

Nets are the wires that connect the components together in the circuit. Depending on the particular language of the netlisting and the features of that language, there may or may not be certain special attributes associated with the nets in a design.

In the analog design environment, the term *schematic* is often preferred. In this environment, a schematic is something designers can see and feel in a *graphic user interface (GUI),* such as the one shown in Figure 4.16, where the primitives are transistors. The circuit block presented by this schematic is an *operational amplifier* (*op-amp* for short), which is a high-gain electronic voltage amplifier with differential inputs and a single output. Op-amps are among the most widely used electronic blocks in analog circuit designs.

In digital designs or the SoC integration environment, the term *netlist* is preferred. This is partly because the design representation is often constructed as a text file, mainly in Verilog netlist format. Figure 4.17 is an example of a very simple digital circuit in which the primitives are logic gates. This

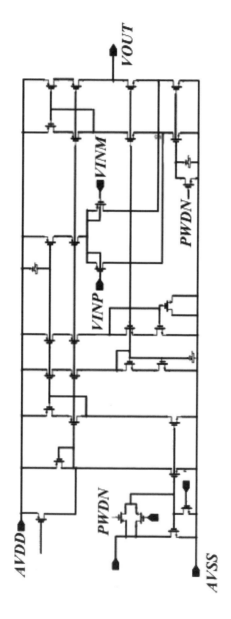

Figure 4.16. A schematic of an analog circuit (an op-amp).

108

```
module DIV10 ( INCLK, OUTCLK );
input INCLK;
output OUTCLK;
wire \counter18[3] , \counter18[2] , \counter18[1] , \counter18[0] ,
\counter[3] , \counter[2] , \counter[1] , \counter[0] , \n89[1] ,
\"+"-return87[3] , \"+"-return87[2] , OUTCLK123, n180, n181, n182,
n183, n184;


AN310 U33 ( .A(\counter[3] ), .B(\counter[0] ), .C(n181), .Y(n180) );
NO211 U34 ( .A(\"+"-return87[2] ), .B(n180), .Y(\counter18[2] ) );
DTN12 OUTCLK_reg ( .CLK(INCLK), .D(OUTCLK123), .Q(OUTCLK) );
NO210 U35 ( .A(\counter[2] ), .B(\counter[1] ), .Y(n181) );
NA210 U36 ( .A(\counter[1] ), .B(\counter[0] ), .Y(n183) );
NO211 U37 ( .A(\"+"-return87[3] ), .B(n180), .Y(\counter18[3] ) );
NO210 U38 ( .A(n183), .B(n182), .Y(n184) );
NO210 U39 ( .A(\counter[0] ), .B(n180), .Y(\counter18[0] ) );
NO211 U40 ( .A(\n89[1] ), .B(n180), .Y(\counter18[1] ) );
IV110 U41 ( .A(\counter[2] ), .Y(n182) );
DTN12 \counter_reg[0] ( .CLK(INCLK), .D(\counter18[0] ), .Q(\counter[0] ));
DTN12 \counter_reg[1] ( .CLK(INCLK), .D(\counter18[1] ), .Q(\counter[1] ));
DTN12 \counter_reg[2] ( .CLK(INCLK), .D(\counter18[2] ), .Q(\counter[2] ));
DTN12 \counter_reg[3] ( .CLK(INCLK), .D(\counter18[3] ), .Q(\counter[3] ));
BF015 U42 ( .A1(\counter[3] ), .B1(\counter[2] ), .C1(\counter[0] ), .C2(\counter[1] ),.Y(OUTCLK123));
EX210 U43 ( .A(\counter[1] ), .B(\counter[0] ), .Y(\n89[1] ) );
EX210 U44 ( .A(n182), .B(n183), .Y(\"+"-return87[2] ) );
EX210 U45 ( .A(\counter[3] ), .B(n184), .Y(\"+"-return87[3] ) );
endmodule
```

Figure 4.17. A netlist of a digital circuit (a frequency divider).

block is designed as a frequency divider of ratio 10. The RTL code of this block is listed in Figure 4.6. Figure 4.17 shows the resultant netlist after the RTL code is synthesized through the logic synthesis step.

In some cases, if the size of a netlist is reasonably small, it can also be visually inspected in a GUI environment. In these cases, this netlist is often called a schematic as well. Figure 4.18 is the schematic view of the same

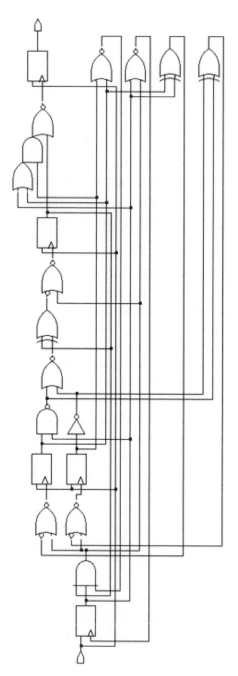

Figure 4.18. A netlist of a digital circuit shown in a GUI (schematic).

netlist of Figure 4.17. In most cases, however, it is impossible to view a digital design's netlist in graphic mode due to its large size.

56. WHAT IS THE GATE COUNT OF A DESIGN?

The term *gate count* is used to measure the size of a digital design. Within the netlist of a digital design, there are many circuit elements or primitives (leaf cells). These circuit elements could be standard cells from an ASIC library or memory or special macro cells. In an ASIC library, different cells have different physical sizes. Memory and macro cells have their unique physical sizes as well. The size of a netlist, or the size of a design, is the sum of the sizes of its components. To conveniently describe the size of a digital design, the physical sizes of the library cells, memory cells, and macro cells are normalized to a base cell, usually a two-input NAND gate with a minimum of 1× drivability (here referred to as NAND2). As a result, the size of any netlist is measured by referencing it to the size of this base cell. The gate count is an estimate of how many equivalent base cells are contained in the netlist. As an example, if a design is a "100 K gates design," its physical size is equivalent to the size of one hundred thousand NAND2 cells.

Do not confuse gate count with the term *number of components.* In the previous example, the actual number of components in the design is not necessarily 100 K since different cells have different sizes. Also, the final actual layout size of this design is not exactly determined by simply multiplying the size of NAND2 by one hundred thousand times. This is due to the fact that extra space is needed for the power grid, I/O busses, clock tree, and routing tracks. In other words, gate count is just an estimate. It is used to conveniently compare the relative sizes of different designs.

Figure 4.19 shows the schematic, symbol, and layout views of a NAND2 cell from an *ASIC* library.

57. WHAT IS THE PURPOSE OF TEST INSERTION DURING LOGIC SYNTHESIS?

As addressed in Chapter 3, Question 20, Design for Testability (DFT) is a very important issue to be considered when creating a profitable commercial product. To enable the DFT function of making the design testable for manufacturing defects, extra circuitry is added in the design, and this extra test circuitry is inserted during the logic synthesis process.

1× drive two-imput NAND gate

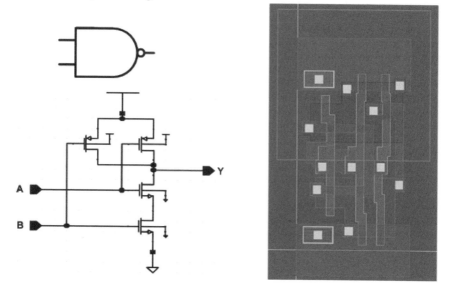

Figure 4.19. Schematic, symbol, and layout views of a NAND2 cell.

If this test circuitry is added in later stages, such as after the place and route stage, rather than during the logic synthesis stage, then the entire physical design process could be jeopardized. This is due to the fact that the test circuitry needs extra silicon area and extra timing budget. Furthermore, the configuration of DFT circuitry has an impact on clock structure and chip/block I/O planning. Thus, test circuitry should be added as early in the chip implementation process as possible.

Some details involved in the insertion of test circuitry are swapping scannable flip-flops, partitioning scan chains, stitching scan chains, constructing built-in self-test (BIST), constructing memory BIST, analyzing fault coverage, and generating test patterns.

58. WHAT IS THE MOST COMMONLY USED MODEL IN VLSI CIRCUIT TESTING?

The most commonly used model in VLSI circuit testing is the *stuck at fault (SAF)* model. The SAF model assumes that any node (a net in a netlist) within a silicon chip has the potential risk of being permanently tied to power (*stuck at one, SA1*) or ground (*stuck at zero, SA0*) due to various manufacturing defects. Either SA1 or SA0 makes the affected node nonfunctional since that node cannot be switched by the circuit for logic operation any longer. Consequently, the chips that contain such nodes are regarded as bad chips and cannot be delivered to the customer. Design for test is the art of inserting some extra testing circuitry inside the chip to search for such SAF nodes.

An SAF is a particular fault model used by fault simulators and *automatic test pattern generation (ATPG)* tools to mimic a manufacturing defect within an integrated circuit. It is not the only model for DFT but it is the simplest and most widely used, especially for digital designs since 0 and 1 are the only concerns in these circuits. However, although very powerful, the SAF model is not a true description of how a node is malmanufactured or damaged. It does not provide physical manifestations of the defect (*defect mechanism*) but only the behavior or effect caused by the defect. For example, metal shorting through a foreign material between a node and ground node or the failure of an internal transistor could both cause an SA0.

Figure 4.20 shows an example of an SA0 fault model. In this circuit, there are seven nodes (or nets): A, B, C, D, E, F, and G. The diagram at the right shows that an SA0 fault is presented at physical location y, which belongs to node G. In other words, node G is always at ground electric potential and cannot be switched by its driver Cell X to logic "1," regardless of Cell X's drive strength. Thus, this circuit is not qualified for its intended design function and should be discarded.

During fault simulation or ATPG, test patterns are generated to stimulate the circuit and detect the effects of such SAFs. During this process, every single node in the circuit is assumed to have the potential of being stuck at either 0 or 1. The aim of a good set of test patterns is to detect all of these faulting nodes in the circuit using a minimum of resources (CPU time, memory, disk space, and tester time, for example).

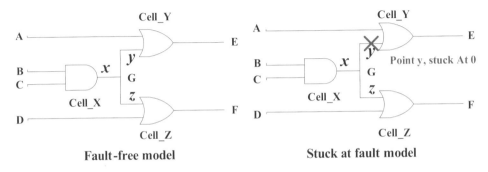

Figure 4.20. Stuck at fault model.

59. WHAT ARE CONTROLLABILITY AND OBSERVABILITY IN A DIGITAL CIRCUIT?

As addressed in Question 58, the most common approach to testing a digital circuit is to toggle every node inside the circuit and observe the corresponding effect. The foundation of this approach is the SAF model. However, in practice, this is not always easily achieved. In a circuit of combinational logic, the logic states of the internal nodes can be determined if the circuit's inputs are all known. But for a circuit that includes sequential elements, such as flip-flops and latches, this is not true. Some of the nodes' logic states depend on these sequential cells' previous states. This leads to *controllability* and *observability* issues.

In the design for testability regime, for any node in a circuit, controllability is defined as the capability of a node being driven to 1 or 0 through a circuit's prime inputs. If this node can be driven faithfully to 1 and 0, it is regarded as controllable. Observability is defined as the capability of the logic state of this node being observed at the circuit's prime outputs. If the logic state of this node can reliably be observed, this node is regarded as observable. Whether a circuit node is stuck at 1 (or 0) is only testable if that node is both controllable and observable.

Figure 4.21 demonstrates the controllability and observability concepts. For a circuit block that has three inputs (A, B, and C) and two outputs (Y and

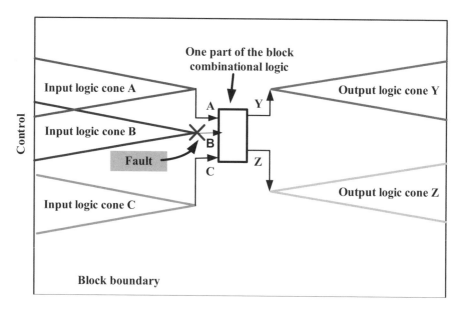

Figure 4.21. Controllability and observability.

Z), node B is controllable when its electrical potential can be brought high or driven low from the prime inputs of this chip through the input logic cone of B. It is observable when the consequence of this switching is sensed by the chip's prime outputs through the output logic cone of Y and Z.

60. WHAT IS A TESTABLE CIRCUIT?

A *testable circuit* is defined as a circuit whose internal nodes of interest can be set to 0 or 1 and in which any change to the desired logic value (the logic value difference between an SAF circuit and a non-SAF circuit) at the node of interest, due to a fault, can be observed externally.

From this definition, it is easy to understand that an *untestable circuit* is the one that has at least one node that is either noncontrollable or nonobservable. There are almost always some noncontrollable or nonobservable nodes in any real circuit of reasonable complexity. Consequently, some untestable faults result. Figure 4.22 shows four types of known untestable faults: the unused fault, tied fault, blocked fault, and redundant fault. These faults cannot be tested by any test vector.

Note: Apparently, untestable faults cannot cause functional failures. Therefore, those faults are not included in the fault coverage calculation.

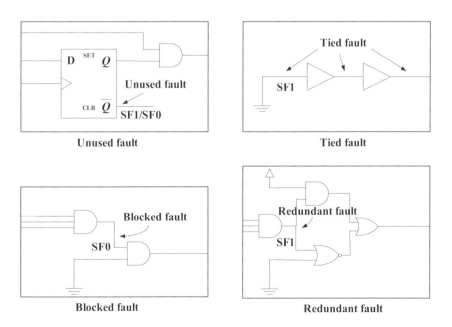

Figure 4.22. Untestable faults.

In addition to the untestable faults, there is a group of *undetectable faults*. These include the *possibly detected fault, oscillatory fault, hypertrophic fault, uninitialized fault,* and *undetected fault*. Basically, these faults include the hard-to-detect or undetected faults that cannot be proven untestable. Unlike untestable faults, which do not have functional impact, undetectable faults must be taken into account when calculating the chip's fault coverage.

In reality, the completely 100% testable circuit is difficult to achieve, especially when inspecting at the chip level.

61. WHAT IS THE AIM OF SCAN INSERTION?

As discussed in Question 60, to create a testable circuit, it is desirable that every node inside the circuit be controllable and observable. However, for a circuit consisting of sequential elements (storage elements such as flip-flops and latches), the controllability and observability cannot be achieved without special care. This is because the logic states of the nodes associated with these sequential elements not only depend on current inputs but also on their previously stored states—they have memory. To determinedly set the nodes to known states, a technique called *scan chain* is usually employed inside the chip.

Figure 4.23 shows part of a scan chain. As depicted, a regular D flip-flop, which has input pins of *D* and *CLK* and output pin *Q*, is replaced by a *scannable* flip-flop that has an extra input pin of *SD*. The idea is to use this *SD* pin to put the flip-flops in known states. Those scannable flip-flops are connected in a chain fashion so that the controlled values are transferred to each flip-flop from the chip's prime inputs through the chain. Also, the logic values of the internal nodes are transferred out to the chip's prime output through the chain.

Figure 4.24 shows an example of using a test vector to detect the SAFs inside a circuit. As shown, a pattern of 000 is applied at three *prime inputs (PIs)*. Also, a 0 is driven to the node of interest through a scan chain. At the

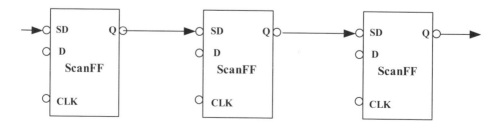

Figure 4.23. Part of a scan chain.

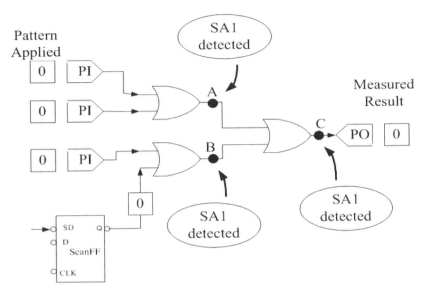

Figure 4.24. A test vector used to detect SAF.

prime output (PO), a 0 is expected if the circuit is fault free. If, instead, a 1 is received at PO, it can be said that one of the nodes A, B, or C is stuck at 1.

The task of scan insertion is tri-fold: replacing the flip-flops with scannable flip-flops, partitioning the scan flip-flops into different scan chains based on certain criteria, and stitching the scannable flip-flops of the same chain using specific methods. The aim of the scan insertion, or DFT, is to make every node within the chip controllable and observable. However, in reality, this is not always achievable. In most design cases, there are some circuit nodes that cannot be controlled or observed, no matter what you do.

In scan insertion, if all the sequential elements in a design are replaced by scannable flip-flops, the approach is called *full-scan DFT.* If only a subset of the sequential elements is replaced due to various reasons, it is referred as *partial-scan DFT.*

In summary, the goal of scan insertion is to improve the circuit's controllability and observability.

62. WHAT IS FAULT COVERAGE? WHAT IS DEFECT PART PER MILLION (DPPM)?

The term *fault coverage* is defined as the ratio of detected nodes (found by test vectors through prime inputs) and the testable nodes inside the circuit.

It is a measure of the confidence level that designers have on the testability of the circuit. If a design bears a fault coverage of 90%, then it simply means that 90% of its nodes are controllable and observable by the current test vector set. Fault coverage of commercial products should reach at least 95%.

In the definition of fault coverage, it is worth noting that it is the ratio between detected nodes and the testable nodes. Thus, in this calculation, it does not include any structurally "untestable" node. In other words, if enough effort is spent on generating the test vector, it is possible to achieve 100% fault coverage. If untestable nodes were also included in the calculation, then, theoretically, 100% fault coverage would be unreachable.

The term *defect part per million (DPPM)* describes the number of parts returned (owing to defects) by customers among one million parts delivered. It is a significant measure of the financial success of a product: the lower the DPPM, the higher the profit margin.

Mathematically, there is no direct relationship between fault coverage and DPPM. However, based on data collected from many products, it has been found that DPPM has an inverse relationship to fault coverage. In other words, a high fault coverage can produce a low DPPM. This is due to the fact that the higher the fault coverage, the more thoroughly the chip can be tested before being delivered to customer and, as a result, the lower the likelihood that the customer will receive bad parts.

Figure 4.25 shows a model between DPPM and fault coverage developed by Williams and Brown. In this figure, DPPM is represented by defect level (scaled by a constant). *Yield* is a major factor in calculating defect level. It is defined as the number of good dies (passing the wafer-level testing) out of one hundred dies. Fault coverage tells us how completely we can test the chip. Yield tells us how many defects occur in the real silicone.

Even if a die passes the production tests, it does not necessarily mean that it is defect free since almost no design can really reach 100% fault coverage and, more importantly, SAF is only a model—it does not cover all of the physical defect mechanisms. Thus, the "good" dies, which pass all of the tests, may still contain defects. However, compared to "bad" dies, which do not pass the tests and are being thrown away, it can safely be said that there is a low possibility that the good dies will contain defects.

A low yield probably means that this manufacturing process is not mature, or is not fine tuned for this design. Consequently, more defects would result in the dies. As shown in Figure 4.25, a product bearing high-fault coverage (design-related) but low yield (mainly manufacturing-related) will still have a relatively high defect level.

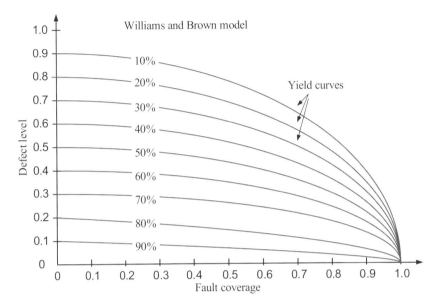

Figure 4.25. DPPM and fault coverage.

63. WHY IS DESIGN FOR TESTABILITY IMPORTANT FOR A PRODUCT'S FINANCIAL SUCCESS?

Design for testability is important because we do not want to deliver bad or malfunctioning parts to our customers. Too many bad parts delivered to customers means financial disaster. Furthermore, it can damage an organization's reputation and result in missed business opportunities.

Without DFT embedded in the design, the only way to differentiate good parts from bad parts is for application engineers or customers to test the chips in the real application environment. In this scenario, the chips are already incorporated in the PCB board or even in the end system (such as a TV or a phone set). A malfunctioning chip identified in this case could mean throwing away the entire system, and the cost associated with that is significantly higher than throwing away a bad chip.

For example, assume that semiconductor company XYZ produces 100,000 chips and sells them to system company ABC for $10 each. Company ABC will use these chips to build PC boards at the cost of $200 each. Eventually, these boards are built into systems that cost $5,000 each. Now assume also that the chips manufactured by company XYZ bear a DPPM of 5%. Thus, out of the 100,000 chips produced, 5,000 are bad. The cost of throwing them away at this stage is $5,000 \cdot \$10 = \$50,000$.

However, if the chips are untested or insufficiently tested and these bad chips are unintentionally used to build the boards, the financial penalty of throwing away (or repairing) these boards is 5,000 · $200 = $1,000,000. Even worse, if these 5,000 bad parts find their way into the final system, the cost of repairing the problem is an astounding number of 5,000 · $5,000 = $25,000,000. This is intolerable; the original selling price for these 100,000 chips is only 100,000 · $10 = $1,000,000. Table 4.1 shows the financial cost of fixing the problem at various stages and different DPPM levels.

As can be seen from this example, the earlier the defective part is identified, the less the financial damage. The lower the DPPM is, the better the chance of financial success. Thus, embedding DFT capability inside a chip and using DFT to test manufacturing defects is crucial for the financial health of a product. Nowadays, it is not surprising to see a chip's testing cost reach more than one-third of the overall chip cost.

64. WHAT IS CHIP POWER USAGE ANALYSIS?

Design for power is a design strategy that addresses the concern of the ever-increasing power consumption by VLSI chips. As process geometries shrink, chip designers can pack more stuff inside a single chip. Consequently, the power consumption by the chip increases correspondingly. To make things even worse, current chips are operating at much higher frequencies compared to chips of ten or twenty years ago. And the power they burn is directly proportional to the operating frequency. Thus, it is not uncommon to see a chip's power consumption exceed tens of watts, or even hundreds of watts. Under these circumstances, a great amount of heat is generated by the chips and cooling is a serious challenge. On another front, more and more mobile applications, which operate on batteries, require low-power operation to achieve longer working hours.

For these reasons, the issue of design for power has moved from backstage to front stage. The formidable challenge posted in this frontier is re-

Table 4.1. The financial cost of fixing problems at various stages at different DPPM levels*

DPPM	No. of bad parts	Chip	PC board	System
5%	5000	$50,000	$1,000,000	$25,000,000
1%	1000	$10,000	$200,000	$5,000,000
0.1%	100	$1,000	$20,000	$500,000
0.01%	10	$100	$2,000	$50,000

*Assumptions: 100,000 chips. Chip cost: $10, board cost: $200, and system cost: $5000.

ducing chip power consumption without significantly degrading chip performance.

One of the key issues in the design for power challenge is the analysis of chip's power usage. Power analysis is the process of analyzing chip or block power consumption based on the circuit topology, the circuit operating speed, the circuit nodes' switch pattern, and the physical layout of the circuit. Ideally, power analysis should be carried out at the transistor level using a transistor-level simulator such as SPICE or the likes. However, for most designs, this is not practical due to the capacity limitation of those simulators. In real design practice, special power analysis tools are available that base their power analysis calculations on power models of standard cells and macros. To obtain reliable power analysis, users must provide the tools with accurate operating frequencies and node switching patterns.

65. WHAT ARE THE MAJOR COMPONENTS OF CMOS POWER CONSUMPTION?

For CMOS circuits, the major components of power dissipation are switching power, short-circuit power, and leakage current. Switching power and short-circuit power are classified as *dynamic power,* whereas power consumed by leakage current is referred to as *static power* (see also Chapter 3, Question 26).

Dynamic power is the power consumed by a circuit when it is active, or when the circuit is performing its logic operations. When the circuit is in an idle state, the dynamic power consumption should be nonexistent. On the other hand, static power is always present even when the circuit is not operating or when its gates are not switching. These three components are illustrated in Figure 4.26.

As can be seen in this figure, when the CMOS circuit (an inverter in this case) is operating (the waveform at pin IN switching from low to high and high to low), it charges and discharges its load capacitor. During this process, the current I_{sw} dissipates power through parasitic resistors as heat. Also, from the waveforms at the middle of this figure, it can be seen that there is a short period during the switching process during which both the p and n transistors are turned ON. During this brief period, there is a direct current I_{sc} flowing from V_{DD} to GND through these transistors. The power associated with this current is called *short-circuit power.*

The I_{lk} current in the figure is the leakage current, which is discussed in Chapter 3, Question 26. Whether the circuit is switching or not, the I_{lk} is always there. In the past, when the process geometries were significantly larger than what we use today and when the power supply voltages were signif-

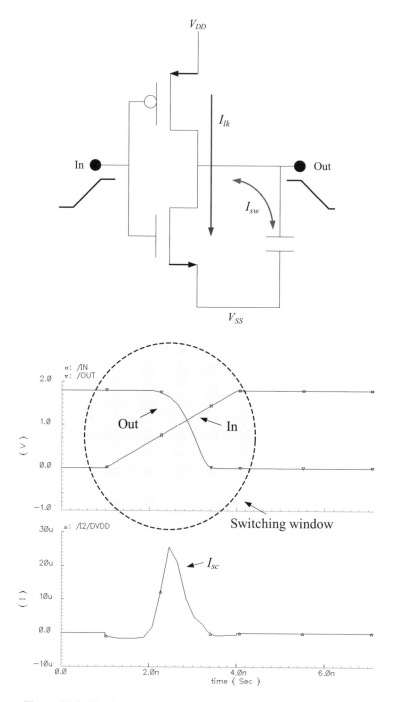

Figure 4.26. The three components of CMOS circuit power consumption.

icantly higher, the power dissipated by leakage current was not of consider-able importance and could be safely ignored. Unfortunately, for today's advanced processes, this is not the case anymore. Controlling leakage current has become a serious and difficult challenge for today's SoC designs.

66. WHAT IS POWER OPTIMIZATION?

Power optimization is a design strategy that aims at reducing the circuit power consumption without significantly degrading circuit performance. A typical example of power optimization is placing the part of the circuit that is not needed at any particular moment into an idle state (the *clock gating technique*). Another commonly used power optimization method is to add extra circuits to control the leakage current, using multiple V_T libraries. A third example is to always run circuits at low operating frequencies unless it is absolutely necessary to run them at higher speeds (this is more or less a system-level technique that is especially useful for general purpose CPU chips).

In the past, during the logic synthesis process, the optimization targets were speed and area. Nowadays, the optimization targets have widened to speed, area, and power. In some cases, power minimization might even be the number one priority.

67. WHAT IS VLSI PHYSICAL DESIGN?

To answer this question in a simple sentence, VLSI physical design includes all the tasks needed to build a silicon chip from design netlist to final GDSII layout.

Compared to design capture, simulation, and logic synthesis (which are called front-end activities), physical design is referred to as back-end work. In other words, physical design covers the tasks necessary to turn the design from a logic entity into a physical entity. It starts from the netlist, which is generated from the logic synthesis. However, the physical design can also include the task of RTL logic synthesis since there is an ever-increasing tie between the logic synthesis and the physical layout. The term *physical synthesis* reflects this fact.

The list below shows roughly the major tasks in the physical design domain:

- Logic synthesis
- DFT insertion

- Electric rules check (ERC) on gate-level netlist
- Floorplan
    - Die size
    - I/O structure
    - Design partition
    - Macro placement
    - Power distribution structure
    - Clock distribution structure
- Preliminary check
    - IR drop
    - ESD
    - EM
- Place and route
- Parasitic extraction and reduction
- SDF generation
- Various checks
- Static timing analysis
- Cross talk analysis
- IR drop analysis
- Electron migration analysis
- Gate oxide integrity check
- ESD/latch-up check
- Efuse check
- Antenna check
- Final layout generation
- Manufacturing rule check, LVS check
- Pattern generation

68. WHAT ARE THE PROBLEMS THAT MAKE VLSI PHYSICAL DESIGN SO CHALLENGING?

As CMOS technology continually scales down to smaller geometries and the density of design integration predictably and ruthlessly scales up according to Moore's Law, IC design productivity must increase at the same rate so that much larger chips are designed within the same man-hours budget. This increase in design productivity must come from new creative design methodologies and advanced EDA tools since the physical structure and mental capability of the human brain has been virtually unchanged for the past hundreds of thousands years. Based on historical evidence to date, design productivity has kept up with technology and design density scaling very well, especially in the front-end design area. However, there are signs of slowdown in back-end productivity. Apparently, this evolution is tougher on physical design than it is on its front-end counterpart.

Besides the design size factor, the difficulties in the physical design regime are due to the issues related to manufacturing technologies, such as device and interconnect parasitics, physical and electrical design rules, device reliability, and process variation. In the practice of real SoC design, these process-related issues surface as the challenges of timing closure, cross talk, electromigration, mask optimization, antenna effect, voltage drop, inductance effect, and chip packaging.

The challenge of timing closure is caused primarily by the parasitic capacitance of the on-chip metal interconnection. In processes of 0.25 μm and below, interconnect wire delay starts to dominate the gate delay on chip timing paths. However, the delay caused by wire parasitics is unknown at the logic synthesis stage. In the past, this problem was dealt with by the *wire load model,* which estimates wire parasitics based on a cell's fan-out and design size. Statistically, this model is reasonably accurate, but the deviation in an individual wire could be off by a large amount. Since timing closure is judged by the worst case of millions timing paths on-chip, this kind of variation is unacceptable. Recently, the concept of *logic effort* has been introduced in design practices to cure this problem. In this approach, during the logic synthesis process, the gate sizes are not fixed as in the conventional approach. Instead, the timing budget, or capacitance load budget, for each gate is determined and fixed, while the gate sizes float. By following this method, the gates sizes are adjusted in a later physical design stage based on real wire parasitics when they are available. This approach provides a better chance for timing closures with faster turnaround times and smaller overall chip sizes.

Cross talk emerges as a major threat to design integrity as process technology continually shrinks. This is due to closer proximities and the taller aspect ratios of the interconnect wires (see Chapter 2, Question 14). Because of this, the area in which they are facing each other increases. Consequently, the capacitance of the parasitics increases as well. This leads to two potential problems: *cross-talk noise* and cross talk delay. The term *cross-talk noise* describes the phenomenon of wrong logic values that are latched into a flip-flop due to the switching of neighboring aggressor wires. Cross talk delay, on the other hand, is the additional delay hit caused by neighboring wires that switch in opposite directions between the aggressor and victim. Cross talk problems can be avoided by shielding and spacing. Additionally, proper gate sizing can significantly reduce the number of aggressors (overdriven nets) and victims (underdriven nets).

When active transistors drive current into wires, the electrons interact with the lattice imperfection. This electron current can be viewed as *electron wind* that slowly moves the atoms. Metal atoms are deposited (as hillocks) over time in the direction of electron flow. In the opposite direction, however, voids grow between grain boundaries. In the long term, voids can reduce wire conductivity, whereas hillocks can introduce mechanical stress or cause shorts with neighboring wires. This is the electromigration (EM) problem in IC design. Controlling current density inside of the wires is the most effective method of harnessing electromigration. Widening the wires and, hence, reducing the current density, can significantly improve *mean time before failure (MTBF)*. For this reason, wide wires are commonly used for a chip's power and clock infrastructure in design practice. The current-handling capability of a via is generally less than that of a metal wire of the same width. Therefore, routers are tuned to insert more vias at layer changes for high-current wires.

In current process technologies, transistor feature sizes are already smaller than the wavelength of light that is used for creating the patterns. As a result, the actual pattern on the silicon deviates significantly from the original GDSII mask data. To correct this distortion, a mask feature called optical proximity correction, or OPC, is added to the mask creation process. Another technique of pushing the lithographical barrier is phase shift masking (PSM), which can help reduce transistor gate length. In the future, using OPC and PSM will become more pervasive.

The antenna problem is introduced during the IC manufacturing process. The IC chip is manufactured layer by layer from the substrate up. After the base layers are finished and the transistors are formed, additional metal layers are deposited for completing the interconnections. During this process, a *logic net* might consist of a number of disconnected metal pieces before the connection is fully completed by all the needed layers. During this period, the static charge on a wire connecting to a silicon gate can result in high voltage and may cause the transistor to break down. This half-assembled wire at an intermediate stage of metal processing can act as an antenna that picks up electric charge. The longer the wire, the more charge it could build up.

A diffusion contact such as a gate output or special-purpose diode could prevent such charge buildup. In practice, to address this issue, antenna design rules are formulated. A typical antenna rule puts a maximum on the ratio between the metal wire length in a layer and the area of the transistor gate. The simplest way to address the antenna problem is to protect each gate input with a diode that has a diffusion contact. However, this results in a large overhead of area and adds parasitic load to the driver. The second approach is to direct the router to jump to higher metal layers near each gate input so that the lengths of the half-assembled wires are short. In this way, the designer avoids the size and load penalties of the diodes.

Voltage drop, or IR drop as it is commonly called, is caused by the resistance of the on-chip network and the electrical current flowing through it. This problem mostly occurs on power networks due to the large currents they must carry. In today's technology, the power supply voltage could be as low as 1 V. The typical IR drop requirement is within 5% of the supply voltage on both V_{DD} and V_{SS}. Assuming that a chip consumes a current of 50 A (50 W of power when V_{DD} = 1 V), this will transfer into 1 mΩ effective resistance requirement for the power network. The sheet resistance is determined by the interconnect material (which will be copper in the foreseeable future) and the thickness of the layer. The layout design tool controls the resistance by modulating the wire width and the number of vias at the intersection of the layer change. In the past, the on-chip power infrastructure was created according to rule of thumb and was oftentimes significantly overdesigned. In future designs with even larger on-chip currents, it will be increasingly more difficult to attack the IR drop problem by overdesigning and not increasing the overall chip size. Therefore, more accurate power simulation and estimation tools are needed to provide a detailed picture of the current distribution on the chip.

Another source of voltage drop is the wire's self-inductance. The slope of the power supply current introduces the voltage drop by $\Delta V = L \cdot (dI/dt)$. Reducing this self-inductance (and thus ΔV) requires the package to have more power pins all over the die. Apart from reducing this self-inductance, the current slope can be trimmed by decoupling capacitors, either on-chip, in the package, or on the board.

The hierarchical approach is a very powerful way to deal with large complicated designs since it scales down the complexity of the problem. It also enables a group of people and computers to work in parallel, which results in faster design convergence. This is especially true for system- and logic-level design, in which there may be many levels of hierarchy. However, this approach is *not* very effective in the physical design domain in which design is often performed in a much flatter fashion, usually with only two levels of hierarchy. The primary difficulty of using multiple levels of physical hierarchy is insufficient automation. Up to now, no feasible block-level, automatic place and route tool exists that can handle blocks of various sizes and shapes while meeting timing requirements and other constraints efficiently. Each hierarchical boundary adds some inefficiency due to the nonideal fit of the modules and the suboptimal placements and connections. It is also harder to hide the parasitic and physical effects, such as cross talk, the antenna, and the wide metal spacing rule, behind the hierarchical abstractions.

In conclusion, it is difficult to deny that VLSI physical design is indeed a very challenging, if not the most challenging, task in the entire chip design process.

69. WHAT IS FLOORPLANNING?

Floorplanning is the first major step in physical design. The key tasks in this step include analyzing the die size, selecting the package, placing the I/Os, placing the macro cells (e.g., memory cells, analog cells, and special-function cells), planning the distribution of power and the clocks, and partitioning the hierarchy.

Die size estimation often starts from the gate count of the netlist (available from the logic synthesis process) plus the physical size of the I/Os and macros. A design can be characterized as I/O limited, core limited, block limited, or package limited. The die size of an I/O-limited design is determined by its number of I/Os. The full placement of the prime input and output cells will dominate the physical size of this chip. On the other hand, in a core-limited design, the size of the chip is governed by the core area or the number of standard and macro cells used. In this case, there is probably room to compensate for a few more I/O signals without increasing the chip size. In a block-limited design, there usually are a significant number of large blocks, or *subchips,* and the chip size is dominated by the sizes of those blocks. For a package-limited design, the chip size is driven by the available package.

Package selection is another major issue that affects the physical design. The selection is based on a number of factors, such as the number of I/Os, the die size, the chip power consumption, and the price. To compensate for the slightly different die sizes, there may be several lead frames available for the same package.

After the package has been fixed, the next crucial step is to arrange the prime input and output cells. I/O configuration has a direct impact on the quality of physical layout since the placement of the rest of the standard cells and macros depend on the I/O locations. The routability of the chip is also closely tied to the I/O configuration. Among many issues, one of the key issues in I/O configuration is the placement of the power and ground pins. These pins, which could amount to up to one-third or more of the total number of I/Os, are placed carefully to reduce or eliminate any *IR* drop or EM problem. Additionally, for complicated SoC chips with many analog macros, there are various special power supplies other than V_{DD} (for core) and V_{DDS} (for I/O). Many of them have to be separated from the main V_{DD}/V_{SS} busses for noise immunization. In such cases, chip I/O planning becomes an even tougher challenge.

Macros such as memories and analog cells are often placed manually by designers based on I/O configuration. Designers must reduce overall routing congestion so as not to create major hurdles for meeting the chip timing target. The placement of those special cells has a great impact on the overall chip placement quality and, consequently, can significantly affect the chip's overall routability. In most cases, it takes several iterations to find good locations for those macros.

In a VLSI chip, every single transistor needs power to perform. The required power is delivered to the transistors through a power distribution network. This network is called the *power plan,* or *power structure,* of the chip. This power network must deliver the appropriate voltage level to the transistors within the chip for their entire lifetime. The two most critical problems associated with a power network are the IR drop and EM. When the effective resistance of the power network is beyond a certain level (such as that caused by narrow metal lines), the voltage drop $(I \cdot R)$ from the source to the destination could be higher than what is tolerable. In such cases, the destination transistors might not function correctly. This is the IR drop problem.

In addition to IR drop, the current flowing through the metal line is constantly pushing and moving the metal atoms. The magnitude of this action is proportional to the current density. After a lengthy period of such action, the metal structure can become damaged, and opens or shorts may result. This is the electromigration (EM) problem. The EM problem will negatively affect a product's life span.

In today's chip operation, almost every action inside the chip is operating on some clock signal. All of the storage elements (flip-flops, latches, and memories) are switched on and off by various clocks. Undoubtedly, the entire chip operation is coordinated by clocks (see Chapter 3, Question 25). Delivering the clock signals reliably to the needed elements is a necessity in physical design. This task is commonly called *clock tree synthesis (CTS).* The two basic concerns in CTS are clock skew and clock tree insertion delay. CTS is a very complicated issue since there are many clock domains in a typical SoC design, and each domain has its own requirement. Sometimes, the clock trees between different clock domains must be balanced as well. Furthermore, in the test mode, the cells in various clock domains must be working at the same testing clock speed. This puts additional constraints on the clock structure.

Another influential issue in floorplanning is hierarchy partition. In some designs, especially large designs, size constraints prevents the entire design from being handled at once by the tools. In such situations, the divide-and-conquer strategy is adopted. A good partition can turn an otherwise unachievable design into a doable one. It can also help speed up the implementation process significantly by enabling parallelism. However, the trade-off is the efficiency of the area and timing. In other words, hierarchical design is not as efficient as flat design in terms of area and timing since the place and route tool cannot see the whole picture at once and consequently cannot perform the optimization as one whole piece.

Figure 4.27 is one floorplan example of a real chip. In this chip, there are almost 400 I/Os, which are located on the chip periphery. About one-third of them are power and ground pins. There are five PLLs, one DLL (delay-locked loop), one high-speed DAC, one large, hard macro on-chip proces-

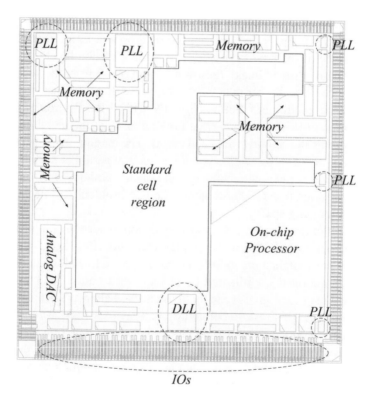

Figure 4.27. An example of a floorplan of a chip.

sor, and more than 80 SRAM memories in this mixed-signal SoC chip. The central area is reserved for standard cells. As apparent in the figure, there are two levels of physical hierarchies: the top level and the standard cell and macro level. As addressed in Question 68, the place and route tool cannot handle the physical hierarchy very efficiently. Having more than two levels of physical hierarchies degrades the quality of the implementation significantly.

The five on-chip PLLs are placed carefully with plenty of space in between. This configuration can effectively reduce interference among the PLLs. The large hard macro is placed in the lower right-hand corner to minimize the impact on the chip's overall routability. The analog DAC is also located in one corner to achieve maximum isolation from the rest of the digital blocks. All of the analog blocks (DAC, PLLs, and DLL) have guard rings embedded in the cell-level layout to minimize noise coupling from the digital circuitry. Moreover, to further trim noise coupling, each analog cell has its own ground, which is not metal-connected to the chip's main digital ground (substrate).

This floorplan is the starting point for the subsequent place and route steps.

70. WHAT IS THE PLACEMENT PROCESS?

Placement is a key step in physical design. As the name suggests, placement is the process of placing the cells, or searching for the appropriate location within the chip floorplan for each cell in the netlist. It is a crucial task because poor placement consumes more area. Moreover, it can impair the chip's speed performance. Poor placement generally leads to a difficult, sometimes impossible, routing task.

A more rigorous definition of the term *placement* follows: given an electrical circuit (a netlist) consisting of a fixed number of components (cells) and the interconnections that describe the interconnecting terminals on the periphery of these cells and on the periphery of the circuit itself, construct a layout indicating the positions of the cells such that the nets are routed and the total layout area is minimized. A further object of high-performance design is minimizing the total delay of the circuit by minimizing the wire length of the critical paths.

The quality of placement work is judged primarily by three factors: the layout area, the completion of the routing, and the circuit performance (timing). There are several algorithms targeted for this placement problem: *simulation-based placement algorithms* (*simulated annealing, simulated evolu-*

tion, and *force-directed placement*), *partitioning-based placement algorithms* (*Breuer's algorithm* and the *terminal propagation algorithm*), and other algorithms (the *cluster growth, quadratic assignment, resistive network optimization,* and *branch-and-bound algorithms*).

Figure 4.28 is a placement example. The placement shown in this figure is a very simple case. There are 129 standard cells and 15 I/O pins in this small design block. The designer first manually places the I/O pins based on his or her understanding of the particular block. Then the placement engine finds the optimal locations for the rest of the standard cells based on the locations of these pins and design constraints. The key optimization targets are meeting timing constraints and minimizing total wire length.

As seen in the figure, the area used for placement is split into horizontal rows, which are referred as *cell rows.* The standard cells in a library are often laid out with the same height, as indicated by the height of the cell row, so that they are placed and aligned together. The small rectangular-shaped blocks are various standard cells (of different widths). An example of one of

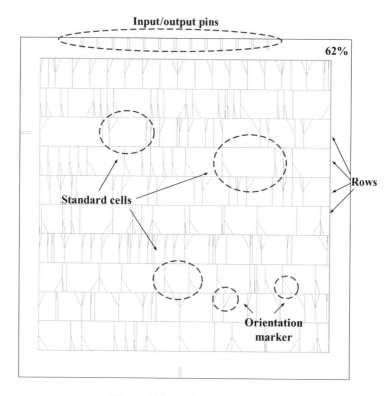

Figure 4.28. A placement example.

those standard cells is a *NAND2* gate, as shown in Figure 4.19. The amount of 62% in the figure is the *utilization rate,* which is calculated as the ratio of the area used by standard cells and the total usable cell row area. From the figure, it seems that the *utilization rate* should be 100% since there is no empty space left in the cell row area. The reason that it is actually 62%, not 100%, is due to the filler cells. There is indeed a certain amount of empty space available after the placement process. However, the empty space is filled with 55 filler cells (the small, narrow cells in the figure) for better manufacturability.

The process of placement is very algorithm intensive and involves great amount of computation work. The amount of computation will increase with the design size at a rate faster than linear. Figure 4.28 is an example of an extremely small block in which the placement work can be finished in less than one minute with any reasonable machine. In contrast, the example shown in Figure 4.27 is a large design. The placement work takes more than 30 hours with a 2 GHz CPU, 16 G byte machine. The largest block (bottom right) in Figure 4.27 is an ARM on-chip processor, which comes into the top level as a layout-finished hard macro. This is the approach of two level physical hierarchies as mentioned in previous text (Question 69). Without this hierarchical partition, the placement process will require even more re-sources (CPU power, memory, etc.). Of course, as also addressed in Question 69, the disadvantage of this physical hierarchy is the sub-optimal solution for the placement, routing, and other physical related issues.

71. WHAT IS THE ROUTING PROCESS?

After the placement step, the exact locations of the cells and their pins are fixed. The next step is to physically complete the interconnections defined in the netlist. This implies using wires (metals) to connect the related terminals within each net. The process of finding the geometric layouts for the nets is called *routing*. As a requirement, the nets are routed within a confined area. Additionally, nets must not be short-circuited. In other words, the nets must not electrically intersect each other.

The objective of the routing process depends on the nature of the design. In general-purpose designs, it is sufficient to minimize the total wire length while completing the connections. In high-performance designs, it is crucial to route each net so that it meets its tight timing target. Special-purpose nets, such as clock, power, and ground nets, require special treatment.

A VLSI chip might contain millions of transistors. As a result, millions of nets need to be routed to complete the layout. In addition, for each net, there may be hundreds of possible routes. Finding the best possible route is very difficult computationally. The routing task has traditionally been divided into two phases. The first phase is called *global routing,* in which a loose route for each net is generated. It assigns a list of routing regions to each net without specifying the actual geometric layout of the wires. The second phase is called *detailed routing,* in which the actual geometric layout of each net within the assigned routing regions is found. Unlike global routing, which examines the layout in the scope of entire chip, detailed routing considers just one region at a time.

Just like the placement process, there are several algorithms to use for global routing: the *maze routing algorithm, line-probe algorithm, shortest-path-based algorithm,* and *Steiner tree-based algorithm.* A number of algorithms exist for detailed routing as well.

Figure 4.29 shows the global route boxes for the placed design of Figure 4.28. The entire layout area is divided into 100 (10 × 10) small regions that are called global routing boxes. The global router's task is to plan the routing configuration in each region such that each cell's pins in this region are assigned a plan for being connected to the appropriate nets. However, the actual metal routing is not accomplished in this step; it is finished in later detailed routing. Figure 4.30 shows the finished routing of this small design block with all of the metal layers shown. Figure 4.31 (page 136) only shows

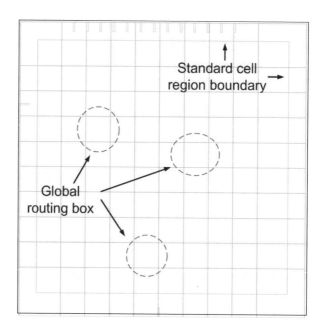

Figure 4.29. Global route boxes.

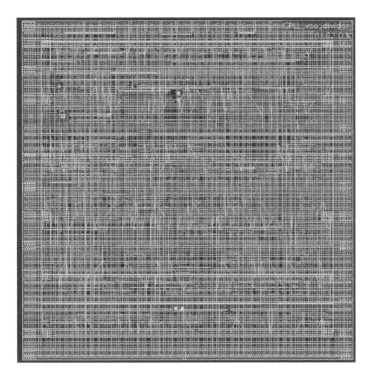

Figure 4.30. The finished detailed route showing all layers.

metal1, metal2, and metal3 for a better view. Figure 4.32 (page 137), which is the zoom-in view of part of Figure 4.30, shows the six levels of metals in detail. Metal1, metal3, and metal5 are used in the horizontal direction, whereas metal2, metal4, and metal 6 are for vertical connections. Also visible are the five layers of vias, via1 to via5, for enabling the metal layer change. Figure 4.33 (page 138) is the detailed route result of a large design with all of the layers shown. The gray background represents the top level metal of metal5. The gray covers the majority of the chip since it is the very top layer. The darker gray is for metal4, which is immediately under metal5 (gray background).

72. WHAT IS A POWER NETWORK?

In order for transistors to perform their assigned tasks, a certain level of voltage must be presented at the transistors' V_{DD} terminals. At the same time, the electrical current traveling through the circuit elements must have a return pass. Thus, the transistors' V_{SS}, or ground, terminals must also be

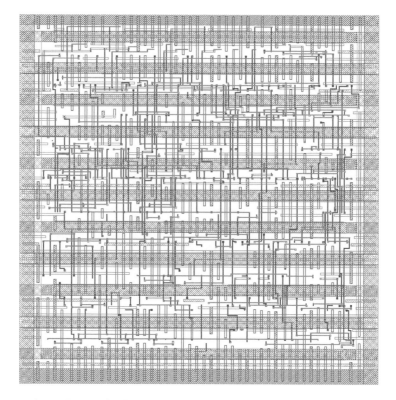

Figure 4.31. A detailed route showing only metal1, metal2, and metal3.

connected. A *power network* is a physical metal structure whose function is to deliver the necessary power (or current) to all of the transistors on the chip. It is the bridge between the power supplies provided outside the chip and the transistors located inside the chip.

A power network is comprised of wide metal lines inside of which current can flow without a substantial resistive force. Figure 4.34 (page 139) is a detailed picture of a power network for the small design in Figures 4.28, 4.29, and 4.30. In this figure, it is seen that the V_{DD} and V_{SS} terminals of the standard cells, which are located on the top and bottom sides of each cell as shown in Figure 4.11, are connected to their dedicated busses. In adjacent rows, the cells are placed in a vertically opposite orientation so that they can share the common V_{DD} and V_{SS} bus.

Figure 4.35 (page 140) is the power network of a large chip. As shown, it is complicated: every macro and memory cell has its own power network, which must be connected to the main chip power network.

As addressed in Questions 19 (Chapter 3) and 68, the main concerns for a power network are IR drop and EM problems. The typical IR drop require-

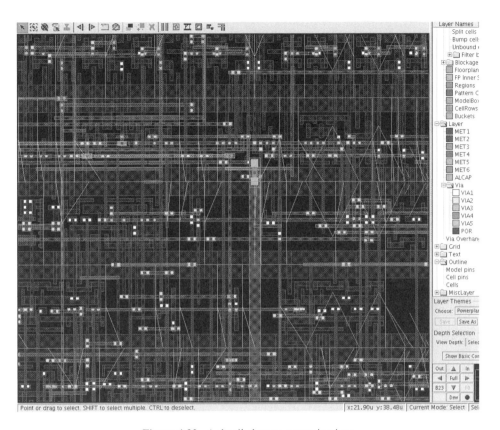

Figure 4.32. A detailed route, zoom-in view.

ment is within 5%. In a 1.1 V power supply design, therefore, V_{DD} cannot fall below 1.05 V, and V_{SS} ground bounce cannot exceed 0.05 V. If the IR drop varies outside this range, there is no guarantee that the circuit will operate properly. The typical method of curing IR drop is to widen the metal lines since wider metal lines result in less resistance. The drawback is an increase in area. Tools are available that can do the IR drop and EM analysis on a power network. Designers must rely on these tools to construct and fine-tune a design's power network for meeting the IR drop and EM requirement without increasing the area too much. This process will likely require several iterations.

An EM check is performed with the calculation of current flowing through each piece of metal on the chip. Then the current densities are checked against the predetermined limit set by the particular process technology. The solution for curing EM violations is also the widening of metal geometry since the current density is reduced when the cross-section area is increased. Unlike the IR drop problem, which has an immediate negative

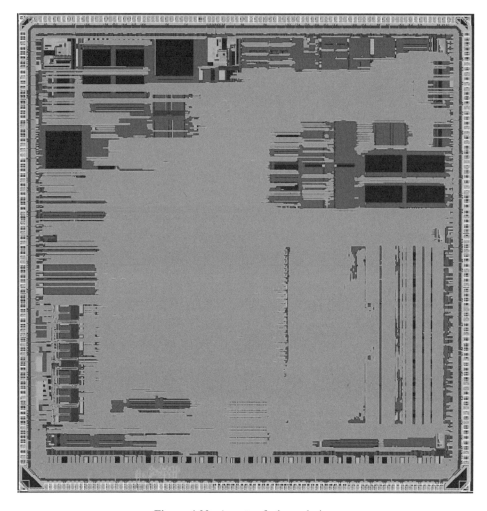

Figure 4.33. A route of a large design.

impact on chip performance, the EM problem is a long-term phenomenon, which will eventually affect the chip's life span.

In mixed-signal SoC integration, the power-planning problem becomes even tougher due to the requirement of separating the analog power supplies from the main digital chip supply. When there are considerable numbers of such analog macros present in the chip, the problem can quickly become an affliction.

Constructing a reliable power network is the most tedious and labor-intensive work in chip physical design. Although there are tools available to ease this pain, it is still mainly stressful manual work.

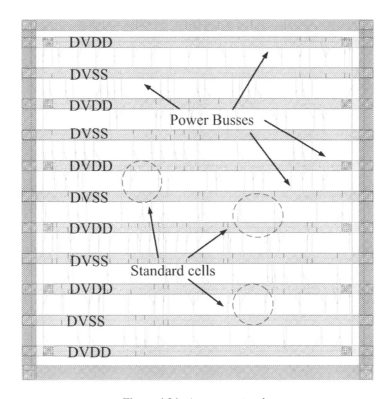

Figure 4.34. A power network.

73. WHAT IS CLOCK DISTRIBUTION?

As addressed in Chapter 3, Question 25, the quality of the clock signals is the most important factor for ensuring a chip's successful operation. In a design netlist, there are hundreds of thousands or millions of cells. Those cells can be classified as two types: combinational cells and sequential cells (including memories). The sequential cells are used for storing information and they must operate on clocks. After the placement stage of the design implementation process, all of the cells, including the sequential cells, are spread around the entire chip. The task of *clock distribution* is to distribute the clock signals to all of these sequential cells. This work is commonly called *clock tree synthesis.*

Figure 4.36 (page 141) shows the principle idea of how a clock tree is constructed. As depicted, a clock network may be constructed in tree fashion. Starting from the clock source, the first level of clock buffers are laid out, then the second level, then the third level, and so on. In most designs, there are many clock domains, and each domain has hundreds or thousands

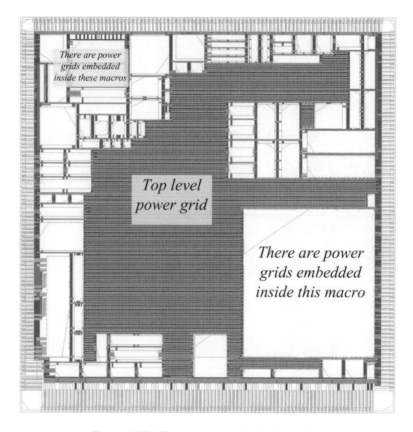

Figure 4.35. The power network of a large chip.

of sequential cells attached to it. This many cells cannot be driven by a single buffer from the clock source, even with the strongest buffer in the library. A tree structure is used to deal with this problem by letting each buffer drive only the number of loads that it is allowed to drive. As a result, the quality of the clock signal, in term of slew rate (the rising and falling time of the clock edges), is not significantly degraded when it reaches the leaf sequential cells. Figure 4.37 shows the commonly used clock tree structures in the clock distribution networks: trunk, branch-tree, mesh, X-tree and H-tree. Figure 4.38 is an example of how a real clock tree looks in a design block. In this simple example, there is one level of clock buffers between the clock root and the leaves.

Another type of clock distribution network is the *clock grid*. In this approach, a grid of metal structure, which covers the entire chip, is dedicated to the distribution of clock signals, as graphically shown in Figure 4.39.

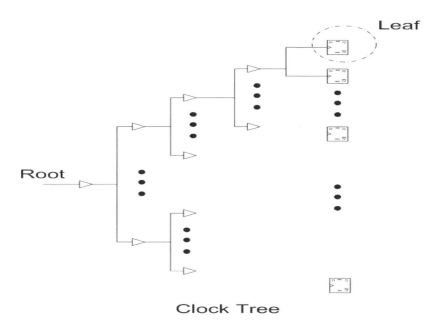

Figure 4.36. A basic clock tree.

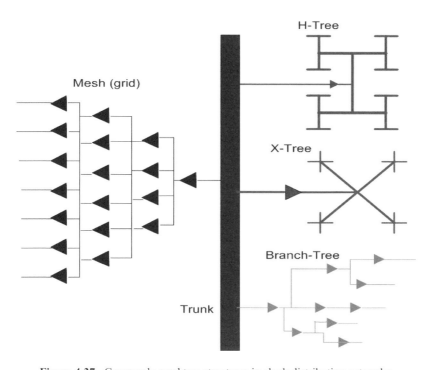

Figure 4.37. Commonly used tree structures in clock distribution networks.

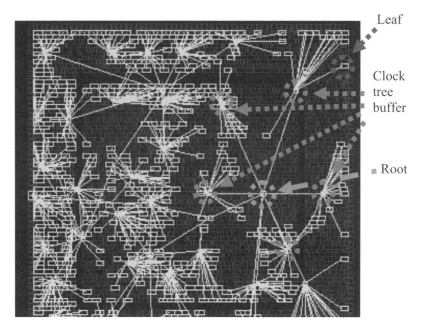

Leaf

Clock
tree
buffer

Root

Figure 4.38. An example of a clock tree in chip design.

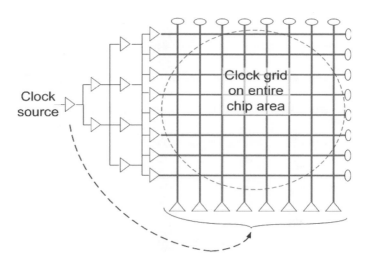

Figure 4.39. The clock grid on a chip.

A tree structure usually consumes less wiring and thus less capacitance and less routing resources, which results in lower power and less latency. However, a tree structure must be carefully tuned and it is very load (placement) dependent. In contrast, a grid structure uses significantly more routing resources and thus has large capacitance and large latency, but it tends to be less load dependent as any leaf cell can always find a nearby tapping point to connect to directly. As a result, a grid structure clock distribution network is typically used only for high-end applications, such as microprocessors, whereas a tree structure is widely used for ASIC-based designs.

The clock distribution network consumes more than 10% of the total power used by the chip in large designs. During each clock cycle, the capacitance associated with the entire clock structure must be charged to the supply voltage and subsequently dumped to ground, with the stored energy lost as heat. To ease this problem, *resonant clock distribution* has been actively studied by some groups. In this method, the traditional tree- or grid-driven clock structure is augmented with on-chip inductors to resonate with the clock capacitance at the clock's fundamental frequency. The energy of the fundamental frequency resonates back and forth between its electric and magnetic form rather than being dissipated as heat. The clock driver is only used for adding the energy lost during the operation. This idea is depicted in Figure 4.40.

74. WHAT ARE THE KEY REQUIREMENTS FOR CONSTRUCTING A CLOCK TREE?

The key requirements for constructing a clock tree are *clock skew* and *insertion delay*. Clock skew is the maximum timing difference among the arrive times of the leaf cells in a clock domain. In Figure 4.41, the result of a SPICE analysis of a clock tree is demonstrated. A clock pulse is injected into the clock tree at time 0 ns with a rise time of 1 ns. After traveling inside the tree, the clock signal arrives at the leaves (also called *clock sinks*) at approximately 3.4 ns. However, it is clear that the arrive times for the leaves are not the same due to the different physical locations of the leaf cells. They spread within a range of approximately 1 ns, which is defined as clock skew. In other words, the existence of skew means that not all of the sequential cells in a particular clock domain receive their clock signals at exactly the same moment, as desired.

Clock skew is significant because it eats up the time budget assigned for logic operations. If skew is over the desired budget, the chip might not func-

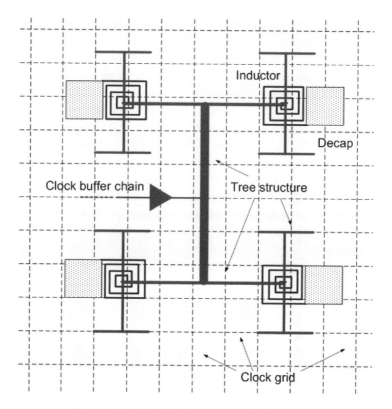

Figure 4.40. The idea of a resonant clock network.

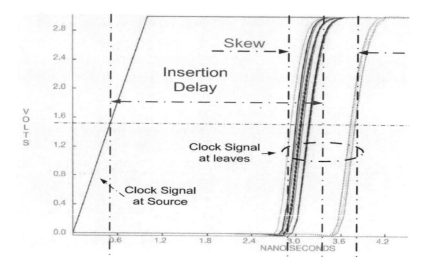

Figure 4.41. Clock skew and insertion delay.

tion correctly at its designed speed (a *setup violation*), or might not function at all (a *hold violation*).

Clock tree insertion delay is the measure of time difference between the clock signal started at the source and the clock signal received at the leaf cells. The concept of insertion delay is also depicted in Figure 4.41. Insertion delay is important because the designer might need to balance clock tree delays between different clock domains for cross-domain information exchange. Also, insertion delay impacts I/O timing constraints. These scenarios are graphically demonstrated in Figure 4.42 where the insertion delays of CLK1_TREE and CLK2_TREE must be balanced for the proper exchange of data between the logic cells of the two domains. For the CLK2 domain, the value of the insertion delay must be known so that the communication between I/O cells (DATA_IN, DATA_OUT) and logic cells can be carried out safely.

75. WHAT IS THE DIFFERENCE BETWEEN TIME SKEW AND LENGTH SKEW IN A CLOCK TREE?

Clock tree synthesis is a crucial step in a chip's physical design. The quality of the clock tree has a great impact on the status of timing closure. One of the critical metrics in measuring the clock tree quality is the time skew, which is the maximum arrive time difference among the clock sinks (see

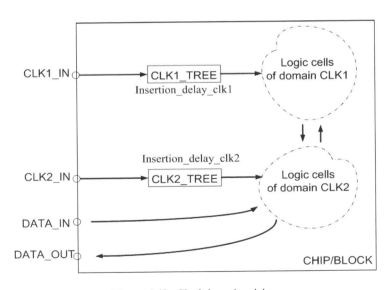

Figure 4.42. Clock insertion delay.

Question 74). Physically, the time skew is caused by the different locations of the clock sinks on the chip. Figure 4.43 is an abstract view of the physical locations of a clock tree's leaf cells. Figure 4.44 presents the same information in a real layout.

As seen, the clock sinks are spread within a certain region. From the clock source to various clock sinks, the physical distances are different. Hence, when connections are completed by metal routing, the wire lengths are not the same. The maximum wire length difference is referred to as *length skew.*

Physically, the clock tree is composed of clock buffers and routing wires. Therefore, the time delay from the clock source to any clock sink is affected by two factors: the gate's delay and the wire's delay. Since these two types of delay scale differently among different *process, temperature, and voltage (PTV)* conditions, a time-balanced clock tree in one PTV corner might experience significant time skew in another PTV corner if the clock tree is constructed with a considerable amount of length skew. This scenario is worsened when the process geometry becomes smaller because wire delay

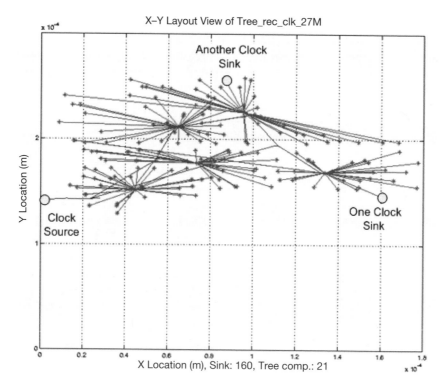

Figure 4.43. Abstract view of the physical distribution of a clock sink.

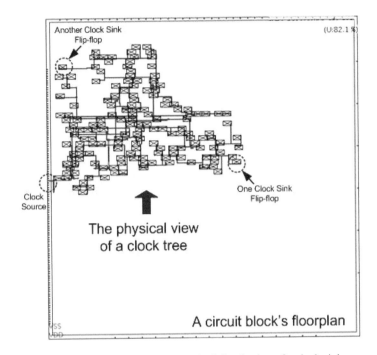

Figure 4.44. Layout view of the physical distribution of a clock sink.

carries more weight in the total delay equation. Ideally, among different branches of a clock tree, it is desired to match gate delay with gate delay and wire delay with wire delay. In other words, time skew should be minimized by using the approach of minimizing the length skew such that the amount of time skew is preserved over different PTV conditions. This is especially helpful for the *on-chip variation (OCV)* optimization.

Figure 4.45 depicts the relationship between time skew and length skew for the same clock tree in Figure 4.44. As shown in this space–timing plot, this clock tree has six levels. Any vertical line in this plot represents a gate delay since a gate has no length skew but time delay. Wire delays are expressed by nearly horizontal lines, which have a large length difference but small time difference. At Level 4 and Level 5, the clock tree starts to grow different branches. Consequently, the length skew is seen at these levels. The time skew for this tree is ~30 ps, whereas the length skew is approximately 250 μm. Figure 4.46 is the same space–time relationship in a three-dimensional (3D) world. Figure 4.47 is the 3D plot of a very large clock tree with 23,942 sinks.

The time skew discussed above is called *global skew,* which is usually pessimistic. A more specific term, *local skew,* is defined as the time differ-

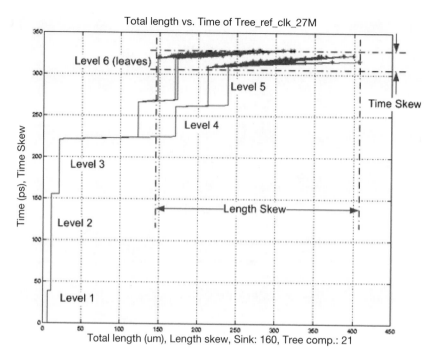

Figure 4.45. The space-time relationship of a clock tree.

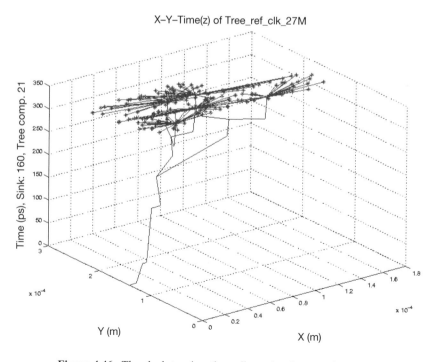

Figure 4.46. The clock tree in a three-dimensional space–time plot.

148

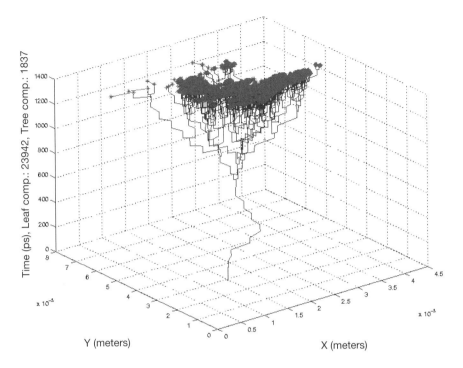

Figure 4.47. A large clock tree of 23,942 sinks.

ence for the clock signals to reach the sinks that have data exchange activities among them. Local skew is more precise and useful for circuit analysis but the extraction of necessary information for processing is beyond the capability of current tools.

76. WHAT IS SCAN CHAIN?

As seen from the answers to Questions 20 and 58-63, design for testability (DFT) is important for making a chip testable. The scan chain is an essential part of DFT strategy that improves a chip's controllability and observability. As shown in Figure 4.23, a scan chain consists of scannable flip-flops stitched together in chain fashion. The intended data is driven into the input port of the scan chain (start of the chain), and the corresponding result of the logic operation (logic value) is observed at the output of the chain (end of the chain). The nodes along the path of the scan chain can be set to an intended value (0 or 1) by the scan chain. The effect of these settings is observed by shifting data out through the chain.

So a scan chain is a chain of shift registers. The data are moved through the chain by the scan (or test) clock. Figure 4.48 is an example of one scan chain in a real design. From the beginning to the end of the chain, several thousands of scan flip-flops are serially connected, as indicated by the *fly-lines*. These flip-flops scatter over a broad area of the chip. Any line in this picture shows the connection between two logically adjacent scan flip-flops. The central area is the region occupied by standard cells. The chain shown in this figure is just one of the several scan chains in the design. There are many more scan flip-flops in this area that belong to other chains.

The logic values of the internal nodes and, thus, the outputs of a circuit that is built with combinational logic can be decided when the logic values of the inputs are set, since combinational logic is deterministic. However, for a sequential circuit, this is not the case, since sequential cells can store information. Thus, the logic states of its internal nodes and outputs depend not only on its inputs but also on the data previously stored in its sequential elements. Using a scan chain, we can set these sequential cells to predetermined states and observe the consequences. The goal is to check whether we can toggle each and every single node inside the circuit to detect SA0 or SA1 nodes inside the circuit. Ideally, all the flip-flops embedded in a design should be replaced by scannable flip-flops so that all the nodes could be

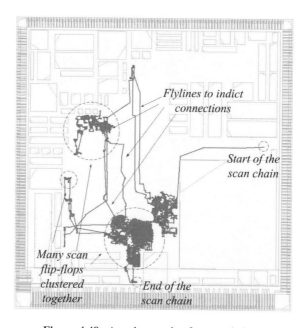

Figure 4.48. A real example of a scan chain.

controlled and observed. However, this is not always possible and is one of the main reasons that test coverage cannot reach 100% in most designs.

77. WHAT IS SCAN CHAIN REORDERING?

As seen in Figure 4.23, a scan chain is a chain of shift registers. In the chain, the output of the previous scan flip-flop is connected to the scan data input pin of the next scan flip-flop. This pattern is repeated from the beginning to the end of the scan chain. Thus, the physical wire connection between any two adjacent flip-flops depends totally on the logical order of the chain. However, logically adjacent scan flip-flops might not be physically adjacent. As a matter of fact, most of time logical proximity does not match physical proximity. Originally, the chain order is decided during the logic synthesis step through a random process or in alphabetical order since the physical locations of the flip-flops are unknown at that time. As a result, if the original chain order is used, the physical wire connections, or routing, are not optimized. A better approach is to disconnect all of the scan chain connections before placement so that the normal placement process is not disturbed by the connectivity embedded in scan chains. Then *reorder* the chain arrangement after the placement step when all of the physical locations of the flip-flops are fixed and known. By reordering the scan chain based on physical information, a much better result is achieved in terms of total connecting wire length and the overall routability of the chip.

Scan chain reordering is a computation-intensive task. An analogy is the *traveling salesman problem (TSP problem),* which to date has no known algorithm for an optimal solution. The problem assumes that a salesman must visit a certain number of cities to promote his products and that every city on the list must be visited once and just once. It is also assumed that there are direct flights between all cities. The question is: What is the best traveling plan that includes all of the cities with a minimum of flight miles? Heuristics algorithms to attack this problem follow:

- *Minimum spanning tree (MST)* algorithm.
- All-MST. Applies MST algorithm using all possible vertexes as starting points. The best one is returned.
- Greedy. Always goes to the closest neighbor until all of the nodes are visited. This is the simplest heuristic.
- Brute force. Performs a brute-force enumeration of all possible tours. In theory, it can find the optimal solution but, in practice, the computation burden is intolerable for any reasonably sized problem.

Figures 4.49 to 4.51 show the results of a one-hundred cities problem with the three algorithms. The Greedy algorithm gives the best result for this problem, with a total distance of 4,541 (an arbitrary unit). All-MST yield 4,972, whereas MST yields 5,217. For a problem in the scale of tens of thousands of cities, as is very common in real scan chain reordering practice, the issue will become very computation intensive, as can be appreciated from these examples.

78. WHAT IS PARASITIC EXTRACTION?

In chip design, the cell instantiations and the logic connectivity are defined in a design netlist. After the cells are placed, the real electrical connections are achieved through metal wires. The cells and the wires have resistance and capacitance that affect the current traveling through them. As a consequence, they impact the signal propagation timing. *Parasitic extraction* is the process of extracting the exact resistance and capacitance values associated with each metal segment so that their impact on signal delay can be precisely determined.

Figure 4.52 is the photo of real metal and via structures in a chip. These metal lines and vias have resistance and capacitance associated with them, and their impact on signal propagation delay must be taken into account in timing analysis. Figure 4.53 is the wire interconnection layout of a small circuit with one inverter driving another inverter. As shown, the layout is com-

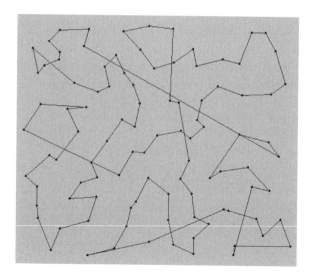

Figure 4.49. The Greedy algorithm: total distance 4,541.

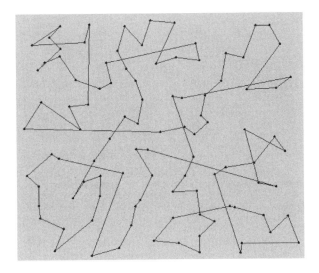

Figure 4.50. The all-MST algorithm, total distance 4,972.

posed of many metal segments. After the parasitic extraction process, the metal connections are represented by an *RC* network.

In the past, when process geometry was much larger than what we are using today, the impact of parasitic *R* and *C* were not as crucial since the signal traveling time was primarily dominated by the cells' delay. However, as

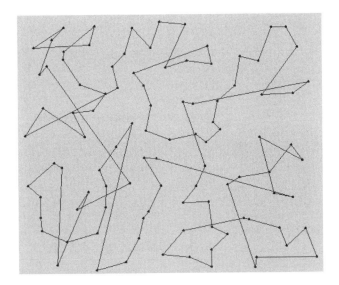

Figure 4.51. The MST algorithm, total distance 5,217.

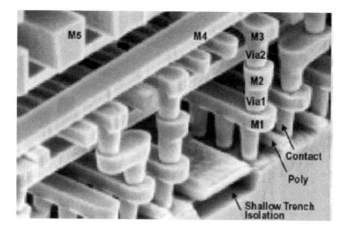

Figure 4.52. Metal and via structures in a chip.

process geometry continually shrinks, parasitic delay has gradually become a dominant factor.

Although the extraction process seems simple and straightforward in this small circuit, full-chip parasitic extraction is very computation intensive and time-consuming. The number of parasitic components on large chips could

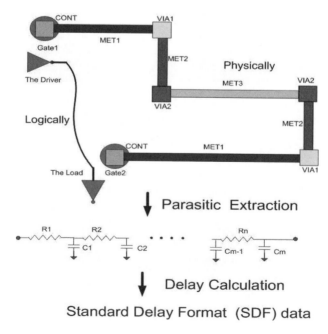

Figure 4.53. The parasitic extraction process.

be huge; for large chips, the extraction results are not suitable for being directly used in timing analysis tools and simulation tools. Therefore, the parasitic extraction process is usually followed by a *parasitic compression* process to compact the information.

The standard parasitic extraction approach today for most chip design is two-dimensional extraction. However, there are more cases where the more accurate three-dimensional extraction is needed, especially for high-performance circuits. The resultant *RC* network from the parasitic extraction process is represented in the *detailed standard parasitic format (DSPF)* or *standard parasitic exchange format (SPEF)*, which can be exchanged seamlessly between CAD tools.

79. WHAT IS DELAY CALCULATION?

After the parasitic extraction process, the interconnecting metal wires are transformed to an *RC* network. Eventually, the impact of this *RC* network on signal propagation delay must be converted to a delay number to determine the operating speed of the chip. This conversion process is called *delay calculation.*

Delay calculation has two parts: the calculation of the delay of a logic gate and the calculation of the delay caused by the wires attached to it. There are many methods for gate delay calculation. The choice depends primarily on the speed and accuracy required:

- Circuit simulators such as SPICE, which are the most accurate, but slowest, method.
- Two-dimensional tables, which are commonly used in applications such as logic synthesis and place and route. These tables use an output load and an input slope to generate a circuit delay and output slope.
- A *k-factor* model is sometimes used, which approximates the delay as a constant plus *k* times the load capacitance.
- *Logical effort* provides a simple delay calculation that accounts for gate sizing. It is analytically tractable.

Similarly there are several ways to calculate the delay of a wire. In the order of increasing accuracy and decreasing speed, the most common approaches are:

- *Lumped C.* The entire wire capacitance is applied to the gate output, which is accounted for in calculating the gate delay. The delay through the wire itself is ignored.

- *Elmore delay* is a simple approximation that is often used where the delay through the wire cannot be ignored and the speed of calculation is important. It uses the *R* and *C* values of the wire segments in a simple calculation. The delay of each wire segment is the *R* of that segment times the downstream *C*. Then all delays are summed from the root.

- *Padé approximations,* also called *moment matching,* are more complex analytic methods. The oldest of such methods is *AWE; PRIMA* and *PVL* are its more recent and sophisticated variants. These methods are faster than circuit simulation and more accurate than Elmore.

- SPICE circuit simulation. This is the most accurate, but slowest, method.

Each of the algorithms has its advantages and disadvantages. The main concerns are accuracy and computation time. The output of the delay calculation process is the data file in *standard delay format (SDF),* which can be used to exchange information among various CAD tools for static timing analysis (STA).

Also worth mentioning is the fact that in the chip design environment, the delay also depends on the behavior of the neighboring nets, which is discussed in Questions 82 and 83.

80. WHAT IS BACK ANNOTATION?

Back annotation is the process of adding the extra delay caused by the parasitic components back into the original timing analysis, which only has the timing from the cells' delay.

Gate-level simulation and static timing analysis (STA) are the two most commonly used approaches in verifying a chip's timing performance. Both of the methods can verify the chip's operating speed against the design specification. In practice, they can be used only with the gate delay information (without the parasitics), as is often done by designers in early design phases. However, in later design stages, when the place and route step is finished and the physical information is available, simulation and timing analysis must be carried out by reading the back-annotated SDF file into the design. Back-annotated static timing analysis is a step that must be performed before the chip can be taped out for manufacturing.

81. WHAT KIND OF SIGNAL INTEGRITY PROBLEMS DO PLACE AND ROUTE TOOLS HANDLE?

The main function of a *place and route (PAR)* tool is to do the cell placement and interconnection wire route routing, as the name place and route implies. However, during the process of performing these functions, the place and route tool also must pay attention to some signal integrity problems, which include cross-talk interference, EM violation, gate oxide integrity (GOI), IR drop, ESD, latch-up, and antenna.

Among these problems, the place and route tool mostly addresses cross-talk avoidance. IR drop and EM problems are studied in great detail mainly during the power network construction step. The IR drop and EM in signal nets have not become a threat to chip functionality yet. GOI check is the process of verifying that none of the transistor gates experience significant voltage overshoot or undershoot for an extended period. If large overshoot or undershoot occurs on any of the transistor gate terminals, it could destroy or damage the transistor and make the chip malfunction.

Protection against electrostatic discharge (ESD) is the technique of protecting those transistors that have direct connection with outside world (through I/O cells). When a chip's pin is touched by a human being or by a device that has static charge on it, the pin could experience a high voltage of 1,000+ V. This high voltage, if not discharged quickly, could destroy the transistors connected to it. ESD protection techniques use specially designed circuits to prevent the damage from happening. They work by quickly discharging the high voltage and large current through other electric paths before damage occurs.

If the diffusion structure in the I/O cells is not designed with care, an unintentional *silicon-controlled rectifier (SCR)* structure could result. The SCR can enter a mode of positive feedback where it is latched on, conducting a large current between power and ground that could destroy the chip. This failure mode is called latch-up.

Antenna failure is caused by long metal wires acting as antennas, collecting electric charge during the manufacturing process in half-assembled metal connections. If the affected metal wires connect to the gate terminal of a transistor, the voltage associated with the charge could damage this transistor.

82. WHAT IS CROSS-TALK DELAY?

By definition, *cross talk* is a phenomenon in which a signal transmitted on one circuit, or one channel of a transmission system, creates an undesired effect in another circuit or channel. Electrically, cross talk is caused by undesired capacitive, inductive, or conductive coupling from one circuit or channel to another.

As shown in Figure 4.54, two electric circuits are very close to each other physically in layout. The metal wires of the two circuits share some com-

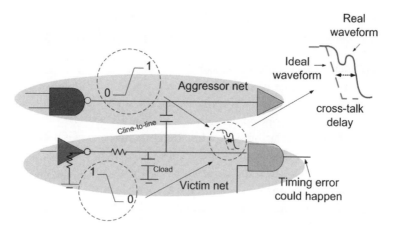

Figure 4.54. Cross-talk delay.

mon silicon area. As a result, the signals traveling through each of the wires can potentially interfere with each other through the coupling capacitor between them. In some cases, the impact of the aggressor (the signal that activates the interference) on the victim (the signal that receives the interference) is the delay addition or subtraction in the victim's signal propagation time. This is called *cross-talk delay*. This deviation of delay could potentially cause a timing error.

83. WHAT IS CROSS-TALK NOISE?

Another effect of cross-talk interference is *cross-talk noise*. Referring to Figure 4.55, the switching of the aggressor signal could cause a glitch on the victim. If the level of the glitch exceeds the noise margin of the receiving gate, the effect of the glitch could pass the gate and result in a logic error and functional failure. Depending on the switching direction of the aggressor net and the current logic value of the victim net, the waveform of the victim net could experience high/low overshoot and high/low noise, as shown in the figure. The high and low noise could potentially cause the logic errors.

Both cross-talk delay and cross-talk noise should be avoided as much as possible for the sake of design integrity. During the place and route step, signals' timing relationships are analyzed. For those signals that have close timing switch windows, the physical space between them in the layout should be increased to reduce the interference. In some cases, stronger buffers can be used in the victim net to boost its signal strength so that less

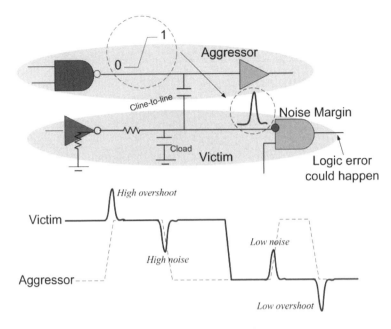

Figure 4.55. Cross-talk noise.

timing damage results. In other cases, replacing the stronger buffers with weaker ones in the aggressor net could achieve the same outcome.

84. WHAT IS IR DROP?

IR drop describes the phenomenon of a drop in voltage potential when a current flows through a resistor of a certain resistance value. In a VLSI chip, the interconnections among the cells are accomplished by metal segments of aluminum or copper. These metal segments are resistors. Thus, the electric voltage potentials at the beginning and the end of the segments are different due to this law: $V = IR$, as shown in Figure 4.56.

IR drop degrades chip performance due to its negative impact on the supply voltage that cells receive. The voltages of the power supplies decrease gradually along the power network as the current flows deeper and deeper into the chip. If, at some locations inside the chip, the amount of voltage deprecation is over a certain limit, the cells at those locations could experience speed degradation, or even worse, stop operating completely.

Figure 4.57 is an example of an *IR* drop plot in a circuit block. The power supply voltage of this block is 1.1 V. The power is provided through the power ports at block's boundary. At the boundary, the supply

A piece of metal with resistance R

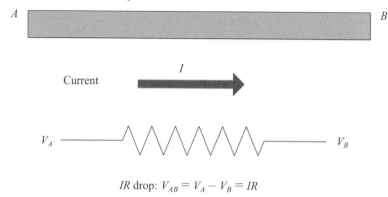

IR drop: $V_{AB} = V_A - V_B = IR$

Figure 4.56. The concept of IR drop.

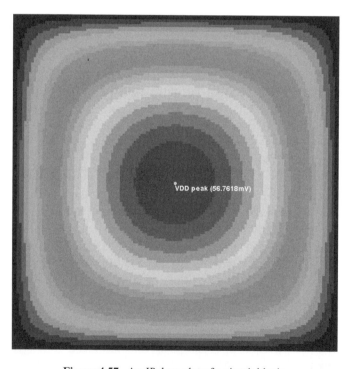

Figure 4.57. An IR drop plot of a circuit block.

voltage is 1.1 V, which is represented by the gray border. Then, gradually moving into the center of the block, the voltage level decreases correspondingly. Most voltage loss occurs at the center of the block. As indicted in the figure, the *IR* drop of the V_{DD} bus at the center is approximately 57 mV or 5% of 1.1 V.

Figure 4.58 is an example of *IR* drop plot for a real chip. The voltage and current sources are at the chip boundary. The current comes into the chip through the power pins that are located at chip's I/O ring. As the current gradually moves into the central region of the chip, the voltage potential drops lower and lower. This is represented by the region of different shades of gray.

IR drop is a serious problem for power nets. It is not as bad a problem for signal nets. This is because for CMOS circuits, the destinations for the signal nets are the gate terminals of transistors, which have high impendence. Therefore, the currents inside the signal nets are likely to be small in magnitude. Consequently, the corresponding *IR* drop is small and can be ignored. On the other hand, the power nets *DVDD* and *DVSS* connect to the source and drain terminals of transistors where significant amount of currents are

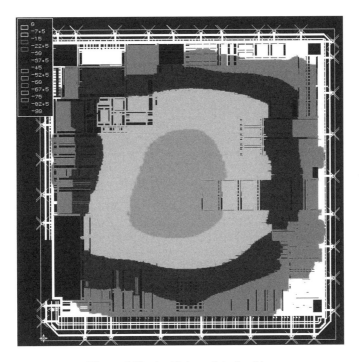

Figure 4.58. An *IR* drop plot of a chip.

present. Also, unlike signal nets, power nets are on a global scale that connects many cells (all of the cells in many cases). Thus the current inside the power net is huge, and the *IR* drop problem cannot be ignored.

85. WHAT ARE THE MAJOR NETLIST FORMATS FOR DESIGN REPRESENTATION?

Logically, the design of a VLSI chip can be completely represented by its netlist. The major formats used for netlisting are:

- Verilog
- VHDL
- EDIF
- DEF
- SPICE

Any of these formats can be used to precisely describe the cells instantiated in the design and the interconnections among those cells. Among those, Verilog is the most popular one.

During project execution, especially for a large SoC project, many tools from different EDA vendors will be used to achieve individual design objectives at specific design stages. Between stages and tools, a Verilog netlist is commonly used for transferring the design information. In the past, it has been extremely difficult to transport a design from one EDA tool to another tool from a different EDA vendor since each company has its own approach to netlisting design. Nowadays, companies seem to standardize their approaches around Verilog.

86. WHAT IS GATE-LEVEL LOGIC VERIFICATION BEFORE TAPEOUT?

As can be appreciated from previous discussions, ASIC design is a complex, multistage, time-consuming process. The associated development cost is so large that it is crucial to have functional silicon at the first attempt. Verification is the process of checking the design's functional correctness. This process can consume over 60% of the total design resource in today's large, complicated SoC designs.

Since design complexity has drastically increased, verification at different stages of ASIC development has become absolutely essential. The designer must verify functionality throughout the design implementation process to minimize the escalated cost of making mistakes at later stages. *Gate-level logic verification* is the last verification step before the design is shipped for manufacturing.

When the design is in its early stages, it is verified by simulation at the system and RTL levels. After logic synthesis, the design presents itself as a gate-level netlist. This netlist will be physically implemented in layout and eventually sent to a manufacturing facility for production. Therefore, the functional correctness of this netlist is very important since this is the logic entity that will be turned into silicon chip. A high degree of confidence in the netlist is required before it can be sent to manufacturing. Gate-level verification is the last chance to check for any functional problems.

Gate-level verification can be performed by gate-level simulation, preferably with back-annotated parasitic components. The verification can also be carried out by emulation, such as a hardware emulation box or a FPGA chip prototype. Another approach is the equivalence check that compares the gate-level netlist with the golden netlist or RTL code.

87. WHAT IS EQUIVALENCE CHECK?

Equivalence check is one of the techniques in a very important verification approach: *formal verification*. Formal verification is the mathematical method in logic verification that thoroughly checks the functional properties of a design. It is an alternative approach to simulation for functional verification. The two major types of formal verification are equivalency checking and formal RTL verification. Equivalence checking is the most successful and easiest to implement formal technique. It can prove that one RTL or gate-level design representation is equivalent to another RTL or gate representation.

For example, during early design phases of a project, the RTL code has been intensively simulated. Much effort has been spent on simulating various cases through sophisticated test benches. As a result, one can be highly confident in this RTL code and it becomes the "golden" RTL that is guaranteed to meet the design's functional requirements. After synthesis, the RTL code is converted to a gate-level netlist. One way of verifying this netlist is gate-level simulation using the previously proven test benches. However, a more efficient way to verify that the netlist is functionally correct is by running an equivalence check. By comparing the logic functionalities of the RTL code and the netlist, one can quickly spot any problem. This concept is depicted in Figure 4.59.

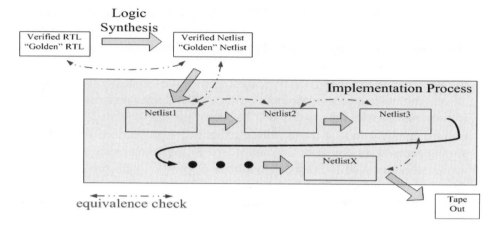

Figure 4.59. Using equivalence check.

As device size and complexity increase, functional simulation becomes more time-consuming both for writing test benches and executing simulation. In contrast, the equivalence check has a short run time and provides complete functional coverage. Subsequently, the equivalence check can be used to compare and verify the design at various levels of abstraction against the golden design.

Equivalence checking has been used intensively in physical design as well. The netlist is altered during many steps of physical design, for example, clock tree synthesis, scan chain reordering, timing optimization, and ECO. Using equivalence check after these steps ensures that a particular step does not alter design functionality. In other words, the primary use of the equivalence check is to establish that successive design iterations still adhere to the functionality of the golden RTL. The application of equivalence checking during physical implementation is demonstrated in Figure 4.59.

Equivalence checking can be performed on the levels of RTL to RTL, RTL to gate, gate to gate, and gate to transistor. Although the history of this technique is not very long, it is proving to be a very useful tool for today's large chip designs and has increasing appeal in the SoC design environment.

88. WHAT IS TIMING VERIFICATION?

Timing verification is the process of checking that the finished chip layout can operate up to the speed defined in the product definition. Unlike functional verification, which addresses chip functionality, timing verification focuses on chip timing or speed. Another difference between the two verifications is that timing checks must be performed in all the process, temperature, and voltage (PTV) corners. A reliable product must run at the designed speed with weak/strong transistors, high/low metal and via resistance and capacitance, high/low temperatures, and high/low power supply voltages. Before the design is sent to manufacturing, the chip's timing behavior under all combinations of these conditions needs to be checked against design constraints.

Timing verification is carried out by either gate-level simulation or static timing analysis. In both cases, the impact of parasitic components must be included in the process. For high-performance special circuits that are often used in SoC integration environments, such as *PLL* (phase lock loop), DAC (digital-to-analog converter), ADC (analog-to-digital converter), the performance evaluation needs to be carried out in more accurate transistor-level SPICE simulation.

89. WHAT IS DESIGN CONSTRAINT?

By definition, a constraint is a restriction on the feasible solutions to an optimization problem. It is any factor that limits the performance of a system with respect to its goal. Designing and manufacturing an IC chip is a multiple-variable-process challenge with various design-related, cost-related, resource-related, and market-related constraints. A technically and financially successful IC project is the production of a good solution within these constraints.

The term *design constraint* refers to the timing requirements imposed on the design by the chip designers. It is used to direct the various steps in VLSI chip implementation, such as logic synthesis, clock tree synthesis, place and route, and static timing analysis. The *standard design constraint (SDC)* format is the standard method of exchanging the design timing requirements among different tools.

For any chip design, design constraint includes three major types of timing requirements: clock definitions, chip/block boundary conditions (input drive strength, input delay, output load, and output delay), and exceptions (false paths and multiple paths). A simple example of a design constraint file is shown in Figure 4.60.

90. WHAT IS STATIC TIMING ANALYSIS (STA)?

Static timing analysis (STA) is a method of computing the expected timing of signals inside a digital circuit without using simulation.

------- Define clock --------------------------

create_clock -period 27.5 -waveform {0 13.75} [get_ports {clkA}]

create_clock -period 13.75 -waveform {0 6.875} [get_ports {clkB}]

create_clock -period 5.5 -waveform {0 2.75} [get_ports {pclk}]

set_clock_uncertainty 0 -setup [get_clocks {clkA}]

set_clock_uncertainty 0.06 -hold [get_clocks {clkA}]

set_clock_uncertainty 0 -setup [get_clocks {pclk}]

set_clock_uncertainty 0.06 -hold [get_clocks {pclk}]

...

------ Define boundary conditions ---------------

set_input_delay 2.5 -clock "clkB" [get_ports {pixel_tol_27M[1]}]

set_input_delay 2.5 -clock "pclk" [get_ports {vs_timing_bypass}]

set_output_delay 3.5 -clock "clkA" [get_ports {vspoldet}]

set_max_delay 1 -from [get_ports {resetb}]

set_max_fanout 20 [current_design]

set_load -pin_load 0 [get_ports {clkA}]

set_driving_cell -lib_cell NA210 -library ABC.db -pin Y [get_ports {resetb}]

set_max_capacitance 24 [get_ports {resetb}]

...

------ Define exceptions ---------------------------

set_false_path -from [get_ports {resetb}]

set_false_path -from [get_ports {actvsovr}] -to [get_ports {vsout}]

Figure 4.60. A simple example of design constraint.

Digital circuits can be characterized by the clock frequency at which they operate. To gauge the ability of a circuit to operate at the specified speed, a designer must measure its delay at numerous development steps, such as at logic synthesis and place and route. Although theoretically such timing measurements can be taken using rigorous circuit simulation, such an approach is too slow to be practical. Static timing analysis plays a vital role in facilitating the quick and reasonably accurate measurement of circuit timing. In static timing analysis, the word static alludes to the fact that this timing analysis is carried out in an input-independent manner. It locates the worst-case delay of the circuit over all possible input combinations. Compared to simulation, the gain in analysis speed is due to simplified delay models. The drawback is the inability to consider the effects of logical inter-actions between signals.

The STA method is not designed for verifying the design's functional correctness, but to check its timing validity. There are huge numbers of log-ic paths inside a chip of complex design. The advantage of STA is that it performs timing analysis on all possible paths. In other words, unlike simu-lation, which only checks timing on given paths, STA is a complete timing check that covers all of the paths, whether they are real or potential false paths. However, it is worth noting that STA is not suitable for all design styles. It has proven efficient only for fully synchronous designs. Since the majority of chip design is synchronous, it has become a mainstay of chip de-sign over the last few decades.

There are several key concepts in STA method: timing path, arrive time, required time, slack and critical path. As shown in Figure 4.61, a *timing path* starts with either a chip prime input or an output pin of a sequential cell (flip-flop, latch, memory). It ends with either a chip output, or the data input pin of a sequential cell. In between, there is logical connectivity such that data can be moved from one point to another. Data is launched either from a chip input or from an output pin of a sequential cell and received at sequen-tial cell's input or chip prime output.

The *arrive time* is the time difference between the moment of launching and the moment the data reaches its destination. It is calculated by summing up the delay amount of each cell along the path. *Required time* is the latest time at which data can arrive without making the clock cycle longer than de-sired. The computation of the required time proceeds as follows. At each primary output or sequential cell's input, the required times for rise/fall are set according to the specifications (for example, clock frequency, or period T) provided to the circuit. Next, a backward topological traverse is carried

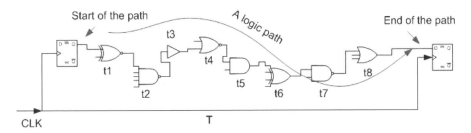

Setup check: t1 + t2 + t3 + t4 + t5 + t6 + t7 + t8 < T

Figure 4.61. A timing path.

out, processing each gate when the required times at its fan-outs are known. The *slack* associated with each connection is the difference between the required time and the arrival time. A *positive slack* of amount Δ at a node implies that the arrival time at that node may be increased by Δ without affecting the overall delay of the circuit. Conversely, *negative slack* implies that a path is too slow, and must be sped up if the whole circuit is to work at the desired speed. The critical path is defined as the path with the worse negative slack.

In STA analysis, based on the concepts presented above, the two checks performed are *set-up check* and *hold check.* Set-up check establishes that the path is fast enough for the desired clock frequency, whereas hold check ensures that the path is not too fast so that data is not passed through. Figure 4.62 demonstrates the set-up and hold concepts. Within a small timing win-

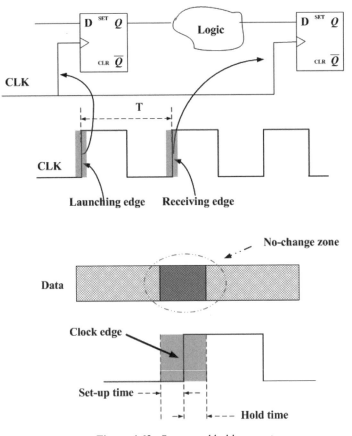

Figure 4.62. Set-up and hold concepts.

dow of active clock edge (circled gray box), the data presented at the flip-flop's data input pin cannot change. Otherwise, the flip-flop will not function as desired. This window is the flip-flop's set-up and hold requirements. This implies that the data lunched from the launching flip-flop must reach the receiving flip-flop before its set-up limit. However, the logic path cannot be constructed too fast. If it is too fast, then within the hold limit of the active edge, the data at the data input pin is already changed. This is the pass through, which is functionally undesired.

The quality of an STA fully depends on the completeness of the design constraints. Figure 4.63 is a sample STA report. In this report, the set-up check on one timing path is shown. The starting point of this logic path is an input pin called "vs_hs_shift_f2[7]." The path ends at the D pins of a flip-flop named "timing_top_inst/inst3/vsout_reg." The clock signal for the set-up check is the signal presented at the flip-flop clock pin "timing_top_inst/inst3/vsout_reg/CLK." This information is given in the first three lines of the report. This particular path meets the required timing constraint with a positive slack of 64 ps, as shown in the fourth line of the report.

The second paragraph of the report lists the components in the clock path. The clock frequency is 182 MHz (5,500 ps) and the reference arrival time (clock tree delay) is 277 ps. The library setup requirement is 115 ps. The margin of *on-chip variation (OCV)* is 46 ps. The last two items eat up time from the clock cycle. Thus, they are presented as negative numbers.

The third paragraph shows that the incoming data will arrive at the D pin of "timing_top_inst/inst3/vsout_reg/D" at the time of 5552 ps. The fourth paragraph shows that the clock signal for this path is "pclk" and the time check's reference point is the rising edge of this clock. The detailed structure of the data path and its individual component delays are shown in the fifth paragraph. The last paragraph is the detailed description of the clock tree structure.

Every path inside the design has a similar report as long as the path is time constrained. Furthermore, the corresponding hold check report is also needed for each path. These reports are valuable during the timing closure debug process.

91. WHAT IS SIMULATION APPROACH ON TIMING VERIFICATION?

As addressed intensively in Questions 42, 43, 44, 45, and 46, simulation is one of the most important methods for chip functional verification. Simula-

```
Start        vs_hs_shift_f2[7]
End          timing_top_inst/inst3/vsout_reg/D
Reference    timing_top_inst/inst3/vsout_reg/CLK
Path slack   64p

Reference arrival time                    277
+ Cycle adjust (pclk:R#1 vs. pclk:R#2)   5500
- OCV Margin (FALL)                       -46
- Setup time                             -115
-------------------------------------    ----
End-of-path required time (ps)           5616

Starting arrival time                      0
+ Data path delay                        5552
-------------------------------------    ----
End-of-path arrival time (ps)            5552

Clock path
pin name                      model name                delay    AT   edge
------------------------      ----------------------    -----   ----  ----
clock:pclk                    sync_processing_afe_top              0   RISE

Data path
pin name                      model name                delay    AT   edge
------------------------      ----------------------    -----   ----  ----
vs_hs_shift_f2[7]             sync_processing_afe_top    2514   2514  FALL
U19/A                         BU130                         0   2514  FALL
U19/Y                         BU130                        90   2603  FALL
BW1_BUF219/A                  BU180                         2   2605  FALL
BW1_BUF219/Y                  BU180                        77   2682  FALL
#top_inst/inst3/U465_C1/A     OR220XSV                      4   2687  FALL
#top_inst/inst3/U465_C1/Y     OR220XSV                    104   2791  FALL
#top_inst/inst3/U131_C1/B     EX220XSV                      0   2791  FALL
#top_inst/inst3/U131_C1/Y     EX220XSV                    118   2909  FALL
#p_inst/inst3/U672_C4_1/B     MU111XSV                      0   2909  FALL
#p_inst/inst3/U672_C4_1/Y     MU111XSV                    121   3030  FALL
#ng_top_inst/inst3/U1_0/A     AD310                         0   3030  FALL
#g_top_inst/inst3/U1_0/CO     AD310                       148   3178  FALL
#_top_inst/inst3/U1_16/CI     AD320                         0   3179  FALL
#_top_inst/inst3/U1_16/CO     AD320                       117   3295  FALL
#_top_inst/inst3/U1_26/CI     AD320                         1   3296  FALL
#_top_inst/inst3/U1_26/CO     AD320                       112   3408  FALL
#_top_inst/inst3/U1_36/CI     AD320                         0   3409  FALL
#_top_inst/inst3/U1_36/CO     AD320                       112   3521  FALL
#_top_inst/inst3/U1_46/CI     AD320                         1   3521  FALL
#_top_inst/inst3/U1_46/CO     AD320                       106   3628  FALL
#_top_inst/inst3/U1_56/CI     AD310                         0   3628  FALL
#_top_inst/inst3/U1_56/CO     AD310                       121   3749  FALL
#_top_inst/inst3/U1_66/CI     AD310                         0   3749  FALL
#_top_inst/inst3/U1_66/CO     AD310                       130   3879  FALL
#_top_inst/inst3/U1_74/CI     AD320                         0   3880  FALL
#_top_inst/inst3/U1_74/CO     AD320                       110   3990  FALL
#g_top_inst/inst3/U1_8/CI     AD310                         0   3990  FALL
#g_top_inst/inst3/U1_8/CO     AD310                       125   4114  FALL
#g_top_inst/inst3/U1_9/CI     AD320                         0   4115  FALL
#g_top_inst/inst3/U1_9/CO     AD320                       109   4224  FALL
```

Figure 4.63. A sample STA report.

```
#g_top_inst/inst3/U1_9/CO      AD320                    109   4224   FALL
#_top_inst/inst3/U1_10/CI      AD310                      0   4224   FALL
#_top_inst/inst3/U1_10/CO      AD310                    109   4333   FALL
#nst/inst3/U1_111_C17_1/B      EX310                      0   4333   FALL
#nst/inst3/U1_111_C17_1/Y      EX310                    166   4499   FALL
#_inst/inst3/U500_C4_1/A2      BH103                      0   4499   FALL
#p_inst/inst3/U500_C4_1/Y      BH103                    102   4601   FALL
#g_top_inst/inst3/U5_C1/B      EX220XSV                   0   4602   FALL
#g_top_inst/inst3/U5_C1/Y      EX220XSV                 131   4732   RISE
#_inst/inst3/U466_C7_7/B3      CM440                      0   4733   RISE
#p_inst/inst3/U466_C7_7/Y      CM440                    166   4899   FALL
#_inst/inst3/U466_C7_10/C      NO420XSV                   0   4899   FALL
#_inst/inst3/U466_C7_10/Y      NO420XSV                 153   5052   RISE
#st/inst3/BW1_INV_H6805/A      IV110                      0   5052   RISE
#st/inst3/BW1_INV_H6805/Y      IV110                     41   5093   FALL
#_inst/inst3/U466_C7_12/C      OR440                      0   5093   FALL
#_inst/inst3/U466_C7_12/Y      OR440                    108   5201   FALL
#inst/inst3/BW1_INV3888/A      IV130B                     0   5201   FALL
#inst/inst3/BW1_INV3888/Y      IV130B                    44   5245   RISE
#_inst/inst3/U467_C3_4/B1      BH055B                     3   5248   RISE
#p_inst/inst3/U467_C3_4/Y      BH055B                    71   5319   RISE
#_inst/inst3/U467_C3_9/B1      BH006                      0   5319   RISE
#p_inst/inst3/U467_C3_9/Y      BH006                     38   5356   FALL
#_inst/inst3/U503_C4_8/A1      BH102                      0   5357   FALL
#p_inst/inst3/U503_C4_8/Y      BH102                    112   5469   FALL
#p_inst/inst3/U503_C4_3/B      MU121                      0   5469   FALL
#p_inst/inst3/U503_C4_3/Y      MU121                     83   5552   FALL
#p_inst/inst3/vsout_reg/D      TIC16                      0   5552   FALL


Reference clock path
pin name                       model name             delay    AT   edge
----------------------         --------------------   -----   ----   ----
clock:pclk                     sync_processing_afe_top           0   RISE
pclk                           sync_processing_afe_top   0       0   RISE
TI_BUFFER_RING_pclk/A          CTB70                     0       0   RISE
TI_BUFFER_RING_pclk/Y          CTB70                    44      44   RISE
clkpclk_atpg_mux/A             CTGMU4                    3       47   RISE
clkpclk_atpg_mux/Y             CTGMU4                  129      176   RISE
clk_pclk_S0/A                  CTI70                     1      177   RISE
clk_pclk_S0/Y                  CTI70                    42      219   FALL
clk_pclk_L0_1/A                CTI70                     4      223   FALL
clk_pclk_L0_1/Y                CTI70                    50      273   RISE
#inst/inst3/vsout_reg/CLK      TIC16                     4      277   RISE
```

Figure 4.63 (*continued*).

tion is also a very powerful technique in verifying a chip's timing characteristics.

During chip development, simulations are carried out at different stages,
such as at the component level (standard cell design, analog cell design, and
memory design), system level, RTL level, gate level, and postlayout gate
level. Accordingly, simulators are divided into the following simulation
modes: behavioral simulation, functional simulation, static timing analysis,

gate-level simulation, switch-level simulation, transistor-level, or circuit-level simulation. Going from high level to low level, simulations progressively become more accurate, but they also emerge as more complex and take longer to run. While it might be possible to perform a behavioral-level simulation of a complete system, it would not be possible to perform a circuit-level simulation of more than a few thousands transistors.

There are several ways to create an imaginary simulation model of a system. One method models the subcomponents of a system as black boxes with inputs and outputs. This type of simulation is called *behavioral simulation* (often using C, VHDL, or Verilog to model). Functional simulation ignores timing. It includes unit delay during simulation, which sets delays of the components to a fixed value (for example, 1 ns). Once a behavioral or functional simulation predicts that a system can work correctly, the next step is to check the timing performance.

At this point, the system might already be turned into a netlist or even a layout. One class of timing simulators analyze logic in a static manner and compute the delay time for each path (see Question 90). This is called static timing analysis because it does not require the creation of stimulus vectors, which is an enormous job for large SoC chips. This type of timing analysis works best with synchronous systems whose maximum operating frequency is determined by the longest path delay between successive flip-flops.

Gate-level simulation is also used to check the timing performance of an ASIC. In a gate-level simulator, a logic gate (NAND, NOR, and so on) is treated as a black box modeled by a function whose variables are the input signals. This function may also model the delay through the logic cell. Furthermore, it can include the impact of the parasitic components by reading in the back-annotated SDF file. Gate-level simulation is mostly used in postsynthesis or postlayout netlists. It is especially useful for verifying asynchronous timing paths since they cannot be handled efficiently by static timing analysis.

If the timing simulation provided by the black-box model of a logic gate is not accurate enough, the next more detailed level of simulation is *switch-level simulation,* which models transistors as on or off switches. Switch-level simulation can provide more accurate timing predictions than gate-level simulation, but the flexibility of logic cells being used as abstraction models is lost. The most accurate, but also the most complex and time-consuming, form of simulation is *transistor-level simulation.* A transistor-level simulator models the transistors, describing their nonlinear voltage and current characteristics.

Simulation is used at many stages during chip design. Initial prelayout simulations include logic-cell delays but no interconnect delays. Estimates of capacitance may be included after completing logic synthesis, but only after physical design is it possible to perform an accurate postlayout, gate-level simulation.

Each type of simulation uses a different software tool. A *mixed-mode simulator* permits different parts of an ASIC simulation to use different simulation modes. For example, a critical part of an ASIC can be simulated at the transistor level while another part is simulated at the functional level. The mixed analog/digital simulators, however, are mixed-level simulators that treat the voltage level of a signal as a variable, not just as 0 and 1 as most simulators do.

92. WHAT IS THE LOGICAL-EFFORT-BASED TIMING CLOSURE APPROACH?

The most critical and challenging issue in SoC implementation is timing closure. The difficulty in timing closure is due to this well-known fact: instead of logic gate delay, the path delay depends more on interconnect delay in current and future process technologies. Traditionally, after logic implementation (from RTL to netlist) the gate sizes in each timing path are determined and fixed before the physical implementation information is available, such as cell locations, interconnect wire length, and so on. As a result, during physical implementation, the synthesis tool will inevitably need to change the sizes of the cells, or even worse, change the configurations of the paths, to compensate for the more dominant interconnect wire delays. Then, after these modifications are made, the new path delays are evaluated and, if necessary, the path configuration is adjusted again. This approach to timing closure requires several iterations to achieve its final timing goal, if it can be achieved at all.

Another not so widely mentioned shortcoming of the current circuit optimization method is that it is not a systematic approach. It is more or less a trial-and-error approach. A simple logic function w = !((!x + y) & z) can be implemented in a variety of logic configurations, such as the three shown in Figure 4.64. For each logic gate (INV, NAND, and OR), there are a certain number of selections available in a specific ASIC library. Even with a very simple library, the number of combinations of different timing configurations for this simple function can be well above thousands. These different configurations must be thoroughly evaluated by the synthesis tool. This trial-and-error approach requires a lot of memory and powerful CPU. For increasingly larger SoC designs, this will be a bigger problem down the road. A consequence of this pseudorandom optimization process is that closure in timing needs a certain degree of luck.

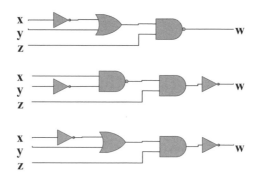

Figure 4.64. Three configurations of w = ! ((!x + y) & z).

These two major drawbacks have become the bottleneck of the current timing closure struggle. For higher-level SoC integration and smaller feature size, a more quantifiable logic optimization approach that reduces or eliminates the randomness in the process is definitely desired.

Designing a circuit of a certain logic function to achieve the greatest speed or to meet delay constraints presents an intricate array of choices. For such typical logic synthesis problems, the following questions have to be answered:

- Which of the several circuit topologies, which produce the same logic function, is the fastest?
- What logic gates' transistor sizes achieve the least delay?
- What is the best number of stages to use for obtaining the least delay?

One approach to answering these questions is trial and error, as mentioned, which is straightforward and well suited for being implemented in simple computer algorithms. The other more systematic approach is realized by the concept of *logic effort*. In their book *Logical Effort—Designing Fast CMOS Circuits,** Sutherland, Sproull, and Harris introduce two key concepts: logic effort and electrical effort. The concept of logic effort is created to quantitatively describe the effort needed to perform a logic function.

The logic effort approach quantitatively measures the effort required for a cell to drive an electrical load. In this approach, logic effort is a measure of how much real estate resource is needed inside a logic gate to perform a spe-

---

Logical Effort—Designing Fast CMOS Circuits, by I. Sutherland, B. Sproull, and D. Harris, Morgan Kaufman Publishers, 1999.

cific logic function. For example, inverting is the simplest logic operation. So it requires the least amount of resources or silicon area. A NAND function is slightly more complicated than inverting. Thus, it requires more silicon resources or a bigger logic effort. Similarly, driving a bigger electrical load requires greater current, which in turn requires more silicon resources. This fact is gauged by electrical effort.

"Nothing is free" is true everywhere and also applies to logic optimization. Faster and more complicated circuit implementation costs more. The concepts of logic and electric efforts were invented to help people to quantitatively measure how much each circuit implementation costs so that different circuit implementations can be compared by using such costs. Ultimately, quick and precise decisions can be made in the selection process of the best-fit circuit configuration for a specific design constraint. This can help to eliminate randomness and uncertainty in the process of logic optimization. Furthermore, it enables designers to develop more intelligent computer algorithms to attack the problems.

In the logic-effort-based circuit optimization approach, there are several key concepts defined for a logic path and for a logic stage.

The key concepts for a given logic stage (a logic gate in a logic path) are:

- Logical effort g
- Electrical effort $h = C_{out}/C_{in}$
- Stage effort $f = g \cdot h$
- Parasitic (fixed) delay p
- Stage delay $d = f + p$
- Delay unit τ

The key concepts for a given logic path that composed of n logic stages:

- Path effort $F = \prod_N f_i$

- Path delay $D = \sum_N f_i$

- Path parasitic delay $P = \sum_N P_i$

- Path total delay $D = D + P$

The logic effort g captures the logic complexity of a given logic gate. It is independent of the sizes of the transistors inside this gate. The logic effort g, which is a unitless number, compares how much worse a logic gate is at producing output current when each of its inputs use the same amount of silicon resource (represented by C_{in}) as that of an inverter. As shown in Figure 4.65, assuming that N mobility is twice that of P, three units of area must be active (to be driven) for an inverter, four units (for any one of the a or b inputs) for a NAND, and five units for an OR to produce one unit of output current. The logic effort g is created to reflect this fact that

$$g \text{ (INV)} = 3/3 = 1$$

$$g \text{ (NAND)} = 4/3 = 1.33$$

$$g \text{ (OR)} = 5/3 = 1.67$$

For any given logic gate, logic complexity is achieved at the expense of silicon resources. This fact is quantified by the introduction of logic effort g. There are also a few theorems using the logic-effort approach:

1. Stage delay $= f + p = g \cdot h + p$ in unit of τ. This is a very significant observation. It is the foundation of quantitative analysis in logic optimization. What is said in this equation is that any given logic gate's delay depends on four contributing factors:
    *   The complexity of the gate function g
    *   The electrical load that this stage drives, h

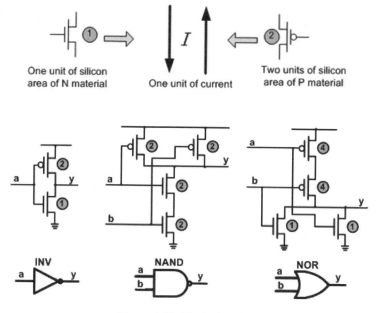

Figure 4.65. The logic effort g.

- An intrinsic delay p that is largely independent of the sizes of the transistors in the logic gate
- The process parameter τ that represents the speed of the process

2. For a given path with fixed number of stages, the path delay is minimized, or the fastest circuit is achieved, when the stages' stage efforts $f = g \cdot h$ along the path are the same, $f = F^{1/N}$.

3. For a given path to achieve the optimal number of stages, the stage effort of each stage shall be $f = g \cdot h \approx 3.6$. And the number of stages shall be $M \approx \log_4 F$. This configuration yields the least path delay or fastest circuit.

These theorems are the tools designers can use in logic optimization. Armed with logic effort g and electric effort h, designers can quickly compute the path delays for the three configurations in Figure 4.64 and determine the best circuit topology for the logic function, without exhaustive analyses and evaluations of all possibilities. Furthermore, since the logic gate delay can be expressed in a linear equation of $d = g \cdot h + p$, analysis and comparison can be carried out without knowing the circuit detail. This improvement in efficiency is enormous for large SoC designs with hundreds of thousands of delay paths.

Logic effort is academic research whose focus is using the least amount of silicon resources to achieve the fastest circuit implementation for a given logic function. SoC timing closure is the real-world practice of designing a chip to meet speed specifications. Timing closure usually starts with design constraints, such as clock speed, input delay, and output delay. Those design constraints are timing budgets for all logic paths in the design. To use the fruit of the logic effort in timing closure, a bridge is needed. The bridge is this brilliant idea: fix the delay budget before the real implementation. This revolutionary idea is only feasible based on the foundation of logic effort.

Using the concepts of logic effort and electric effort, the delay of each logic stage can be quantitatively expressed as $f = g \cdot h + p$. So the timing closure can be planned by analysis before the physical design even starts. Before the physical implementation, the analysis for each path based on logic effort g and electric effort h is already a good indicator of whether the design can be implemented or not, which helps avoid costly mistakes. In the downstream implementation, the task is to simply select the actual cell for each stage based on its g and h. This is the "fix time first, vary size later" philosophy. The traditional approach is characterized as "fix cell first, then evaluate timing, then fix cell again, then evaluate timing again. . . ." This approach indicates the inevitable need for iterations between logic implementation and physical implementation. The logic effort approach avoids this iteration by its nature since timing closure's ultimate goal is timing closure, not size closure.

In summary, the key advantages of this revolutionary logic-effort-based timing closure methodology are:

- Quantified analysis has been brought into the process of logic optimization, which makes more intelligent computer algorithms possible, which in turn makes the process less CPU and memory intensive compared to "trial-and-error" approach.
- Decent analysis can be performed before the actual implementation, which can help reduce the possibility of costly mistakes.
- In theory, using this approach, it is possible to implement designs just as needed. Overdesign, in term of area and power, can be avoided.
- By its nature, no iteration between logic implementation and physical implementation is needed.

93. WHAT IS PHYSICAL VERIFICATION?

Physical verification is the process of checking that the finished layout complies with the manufacturing rules associated with this process and agrees with the schematic/netlist.

Physical verification, as its name suggests, focuses on the physical aspects of the design. This verification process has nothing to do with the timing or logic aspects of the design. The logic aspects of the design are verified by simulation or hardware emulation. The timing aspect of the design is guaranteed by gate-level simulation with back-annotated parasitic delay or by static timing analysis.

In physical verification, a set of design rules check the design's manufacturability. If design rules are violated, the design may not function. Design rules are a series of parameters provided by semiconductor manufacturers that enable the designer to verify the physical correctness of the design. Design rules are specific to a particular semiconductor manufacturing process. A design rule set specifies certain geometric and connectivity restrictions to ensure sufficient margins to account for variability in semiconductor manufacturing processes, so as to ensure that most of the parts operate correctly.

Academic design rules are often specified in terms of a scalable parameter, λ, so the geometric tolerances in a design may be defined as integer multiples of λ. This simplifies the migration of existing chip layouts to newer processes. In contrast, *industrial rules* are highly optimized, and only approximate uniform scaling. Design rule sets have become increasingly more complex with each subsequent generation of the semiconductor process.

The main objective of physical verification is to achieve a high overall yield and reliability for the design. To improve die yields, physical verification has evolved from simple measurement and Boolean checks to more involved rules that modify existing features, insert new features, and check the entire design for process limitations such as layer density, antenna violation, and via overhang. Nowadays, a completed layout consists not only of the geometric representation of the design, but also the data that provide support for manufacturing the design. Although design rule checks do not validate that the design will operate correctly, they are constructed to verify that the structure meets the process constraints for a given design type and process technology.

Physical verification software takes a layout in the GDSII standard format as input and produces a report of design rule violations that the designer may or may not choose to correct. Design rules are sometimes carefully stretched or waived to increase performance and boost component density at the expense of yield. In some cases, the waiving of certain rule violations is simply due to schedule pressure. However, in these cases, the consequence must be understood and taken into account in the financial management of the project.

Physical verification is a computationally intense task. Unlike in some of the previous implementation steps, at this stage all the detailed information is included in the design database since this is the final layout that will be sent to fabrication. If a large verification job is run on a single CPU, engineers may have to wait up to a week before the result returns from a check run. In most designs, the check must be run several times prior to completing the design. Therefore, parallelism with the capability of using multiple CPUs is usually incorporated in the physical verification software.

94. WHAT ARE DESIGN RULE CHECK (DRC), DESIGN VERIFICATION (DV), AND GEOMETRY VERIFICATION (GV)?

Physical verification has two parts: DRC check and schematic verification. DRC stands for design rules check. The term DV stands for design rule verification and GV stands for geometry verification. All the three terms point to the same process of checking that the layout is compliant with manufacture rules.

The basic circuit components such as P and N transistors, capacitors, and resistors are built using various process steps: oxidation, diffusion, ion implantation, chemical vapor deposition, metallization, and plasma etch. The size of the components is controlled by how the designers draw these layers in a layout, which is a representation of the actual process steps. Just as manufacturing must follow certain rules when the components are built using these process steps, so must the layout designers honor them. For example, the following rules apply to almost all CMOS processes:

- Active-to-active spacing
- Well-to-well spacing
- Minimum channel length of the transistor
- Minimum metal width
- Metal-to-metal spacing
- Metal fill density

Figure 4.66 is the complete picture of metal1 rules for a particular process. As shown, there are many checks in just this one metal layer. Rules A and B specify the metal width, Rule K is for metal spacing, whereas Rule O is for wide metal spacing. In modern process technology, the complete

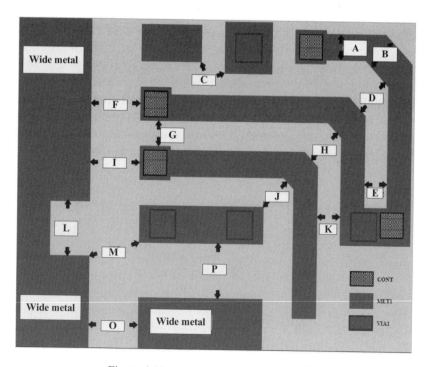

Figure 4.66. A complete rule set for metal1.

design rule set contains many more checks for layers similar to this. Some are more complicated than the example shown here and some are simpler. During the lifetime of a process, the design rule set is constantly updated as the process gradually matures. Requiring a DRC-clean layout helps ensure that the chip being manufactured has a high possibility of success. If a design does not bear 100% DRC-clean stamp, it does not necessarily mean that all the chips will fail functionally, but that the design does have a high risk of *yield* loss.

95. WHAT IS SCHEMATIC VERIFICATION (SV) OR LAYOUT VERSUS SCHEMATIC (LVS)?

The other part of physical verification is the schematic/netlist check. It is referred to as *schematic verification (SV)* or *layout versus schematic (LVS)*. Both terms describe the process of validating that the layout matches the netlist/schematic.

A chip layout consists of millions of geometries of many layers. Certain types of extraction tools are needed to extract the circuit components from these pure geometries. These extraction tools depend solely on specific rules to compose the circuit components from those polygons. Those rules instruct the extraction tool on how to recognize P and N transistors, capacitors, and resistors from the geometries' configuration. SV or LVS are these extraction processes, and they ensure that the circuit resulting from the layout agrees with the netlist/schematic.

For example, Figure 4.67 is a schematic. This small circuit block cleans up the lock-detect signal from a PLL. Figure 4.68 shows the layout of this block. Based on the given rules, the extraction software extracts a circuit from the layout, which should match the circuit in Figure 4.67.

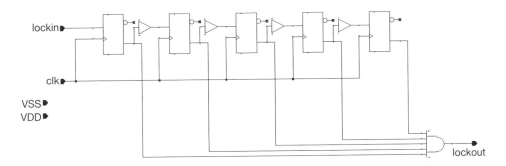

Figure 4.67. A circuit schematic.

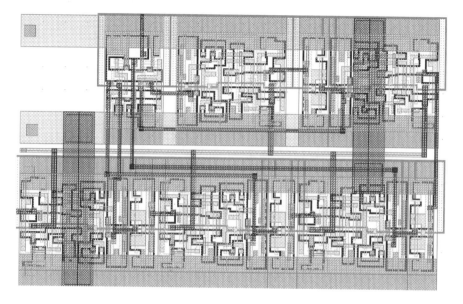

Figure 4.68. A circuit layout.

SV errors (in which the schematic/netlist does not match the layout) can be very difficult to debug. This is especially true for large SoC designs in which there are millions of components on board. It is extremely important for each component used in the SoC to be SV clean at the cell level. Otherwise, the top-level SV task would become a disaster.

Unlike the DRC check, which might allow some unresolved violations, being SV or LVS clean (free of violations) is an absolutely necessary condition before a design can be sent to manufacturing.

96. WHAT IS AUTOMATIC TEST PATTERN GENERATION (ATPG)?

From Question 58, it is seen that a fault model is a hypothesis of how a circuit may fail during the manufacturing process. A fault is said to be detected by a test pattern if, when applying the pattern to the circuit through primary inputs, different logic values between the original circuit and the faulty circuit are observed in the circuit's primary outputs. *Automatic test pattern generation (ATPG)* is an automated process that attempts to find an input sequence (test pattern) enabling the tester to distinguish between the correct circuit behavior and the faulty circuit behavior caused by a particular fault.

ATPG consists of two phases for a given target fault: *fault activation* and *fault propagation*. Through primary inputs at the fault site, fault activation establishes a signal value opposite to that produced by the fault. Fault propagation propagates the fault effect forward by sensitizing a path from the fault site to the primary outputs.

The most popular fault model used in practice is the single stuck-at-fault model (see Question 58). In this model, at a given moment, it is assumed that one of the nodes in a circuit is stuck at a fixed logic value, regardless of what inputs are supplied to the circuit. The stuck-at-fault model is a logical fault model since no timing information is associated with the fault definition. In contrast to intermittent and transient faults that only appear randomly through time, the stuck-at-fault model is also called a permanent fault model because it is assumed that the faulty effect is permanent.

A pattern set with 100% stuck-at-fault coverage must consist of test vectors that can detect every possible stuck at fault in a circuit. However, a 100% stuck-at-fault coverage does not necessarily guarantee high test quality since many other kinds of faults (such as bridging faults or opens) often occur as well, which cannot be modeled by this simple stuck at behavior.

Several algorithms have been used during the evolution of ATPG technology: the D algorithm, path-oriented decision-making (PODEM) algorithm, fan-out oriented (FAN) algorithm, and so on. The D algorithm was the first practical test generation algorithm in terms of memory requirements. The D notation introduced in the D algorithm is continually used in later, more advanced ATPG algorithms. The path-oriented decision-making (PODEM) algorithm is an improvement over the D algorithm. It was created in 1981 when shortcomings in the D algorithm made it evident that design innovations had resulted in circuits that could not be handled by the D algorithm. The fan-out oriented (FAN) algorithm is an improvement over the PODEM algorithm. It limits the ATPG search space to reduce computation time and accelerates backtracking. Methods based on "satisfiability" are sometimes used to generate test vectors as well. *Pseudorandom test generation* is the simplest method of creating tests. It uses a pseudorandom number generator to generate test vectors.

ATPG efficiency is another important consideration. The effectiveness of ATPG is measured by the fault coverage achieved for the fault model and the number of vectors needed, which is directly proportional to the chip's test time or test cost. ATPG efficiency is influenced by several factors, such as the fault model under consideration, the type of circuit under test (full scan, synchronous sequential, or asynchronous sequential), and the level of abstraction used to represent the circuit under test (gate, transistor, or switch).

97. WHAT IS TAPEOUT?

Tapeout is the final step of chip design. It is the time at which the design is fully qualified and ready for manufacturing. After the physical design is finished, the functionality of the netlist is verified, and the timing analysis is satisfied, the final layout, usually in GDSII format, is sent to mask shop to generate photomask reticles. The resultant masks will be used to direct the manufacture of this chip.

The term tapeout originally referred to the action of writing the final data file that describes the circuit layout onto magnetic tape. This term is still used today even though magnetic tapes are now rarely used for this process. More precisely, this process should be called pattern generation (PG). As we know, semiconductor device fabrication is a multiple-step sequence of photographic and chemical processing in which electronic circuits are gradually created on a wafer made of pure semiconductor material. Each of the steps require photomasks, which are created during the pattern generation process, to guide the operation. In this regard, PG is a more appropriate term.

Tapeout is a major milestone during a product's development. It is often succeeded by a celebration among the people who worked on the project, followed by eager anticipation of the actual product returning from the manufacturing facility.

98. WHAT IS YIELD?

After being fabricated, the semiconductor devices are subjected to a variety of electrical tests to determine if they can function properly. In general, the fraction of devices that are capable of performing properly is referred to as the *yield*.

In detail, there are three types of yield:

- Process yield (Y_P). The percentage of wafers that survive the manufacturing process.
- Test yield (Y_T). The percentage of dies that pass the electrical test.
- Assembly yield (Y_A). The percentage of devices that pass the packaging assembly line.

The overall yield is given by $Y = Y_P \cdot Y_T \cdot Y_A$.

There are other slightly different yield terms that basically express the same concept:

- Line yield. The fraction of wafers that are not discarded before reaching the wafer electrical test.
- Die yield. The fraction of dies on a yielding wafer that are not discarded before reaching the assembly line.
- Final test yield. The fraction of devices built with yielding dies that are acceptable for shipment.

The overall yield is the product of these three terms. Manufacturing facilities have direct control over line yield and die yield. However, they only have influence on the final test yield.

Yield loss is also classified as *functional yield loss* and *parametric yield loss*. Functional yield loss consists of dies that do not function, whereas parametric yield loss refers to the dies that do function, but not according to the specification (they may have lower frequency, incompatible voltage range, etc.).

The sources of yield loss are categorized as the result of either random or nonrandom defects. Random defects are typically induced by particles that create pinholes in the resist or in material to be deposited or removed, or by particles that block exposure of the resist. Nonrandom defects include scratches, wipeout, global misalignment, overetching, large-area dislocation, and line-width variations across the chip. In the early stages of a process, nonrandom systematic defects are prevalent. As the process gradually matures, these defects are rapidly eliminated or minimized, and random defects become dominant.

The sources causing defects are classified into four groups: human, environment, equipment, and process. In modern processes, equipment and process are the main sources of random defects. Figure 4.69 shows some ex-

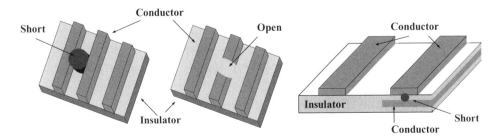

Figure 4.69. Examples of defects in the manufacturing process.

amples of defects that occur during the manufacturing process. The left and middle cases are defects occurring on one layer. The left case shows a short between two conductors (metal pieces) by an extra piece of foreign material lodged between them. The middle case is an open caused by missing material. The case on the right is also a short caused by a foreign material, but it occurs between two different layers.

The ability to predict yield before a chip's full production is highly desirable since it enables engineers to take corrective actions in the manufacturing process. The advantages of defect-mechanism study and yield modeling include identifying the causes of defect types for a particular chip or a range of chips, showing which defect types cause the most yield loss, identifying when a fabrication process is not performing as expected, determining the extent of parametric problems (design and process), monitoring the fabrication process, identifying chip design problems that result in yield loss, driving layout modifications to enhance yield, and identifying the best design practice for a particular process.

An accurate yield model requires both the chip's critical areas and the process defect data. The critical area is defined as the area of die that will kill the functionality of the die if a defect of given size lands on it. The procedure typically involved in a yield prediction study is: (1) calculating the critical area, (2) modeling the defect size distribution, and (3) combining this information to estimate yield loss. The most widely used yield model is the *Poisson model:*

$$Y_{\text{Total}} = Y_0 \prod_{j=1}^{k} e^{-\lambda_j}$$

where Y_0 is a factor that acts as an adjustment for any nonrandom defects and λ_j is the average number of faults of type j.

This learning process based on yield model and defect data is critical in cutting the time from development to market and reducing production cost. Typically, the yield curve follows the pattern as illustrated in Figure 4.70.

During the first stage, initial flaws in the design and manufacturing process are identified and corrective actions are implemented, primarily to eliminate the systematic yield detractors. During the second stage, rapid yield learning results in reducing both random and nonrandom defect density; yield improvement occurs quickly in this stage. During the third stage, the remaining random defects become the dominant yield detractors and the rate of improvement decreases significantly.

Overall, yield is a very important factor in determining the chip's final selling price or profit margin: the higher the yield, the greater the profit margin.

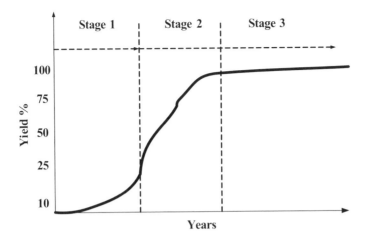

Figure 4.70. Typical progression of yield learning.

99. WHAT ARE THE QUALITIES OF A GOOD IC IMPLEMENTATION DESIGNER?

This book is entitled *VLSI Circuit Methodology Demystified*. It primarily addresses the group of engineers whose job responsibility is chip integration. The focus point of chip integration is the implementation, not the circuit design. Unlike transistor-level circuit designers who spend most of their time on the architecture, analysis, optimization, and simulation of small circuit components, chip integration engineers (or implementation engineers) mostly work on the task of turning a large chip design from a logic entity (RTL description or netlist) into a physical entity. The spirit embedded in this activity is "put everything together and make it work," not "create/invent something from scratch."

Due to the nature of chip integration, the characteristics of a good IC implementation engineer are:

- Be widely knowledgeable in the areas of semiconductor processing, RTL coding, simulation, logic synthesis, place and route, custom layout, and VLSI circuit testing. The key descriptor for the knowledge base is "wide," not necessarily "deep."
- Be familiar (have hands-on experience) with the EDA tools in at least three areas: simulation, synthesis, and place and route.
- Be familiar (have hands-on experience) with some type of ASIC design flow.

- Be skilled in programming. Be familiar with the script languages, such as "tcl," "perl," and "awk," that are used widely in the chip integration environment.

- Be good at solving problems, instead of focusing on just being a tool expert. S/he also must establish an "engineer sense," that is, knowing "what is important at what time." The work of IC implementation, just like any other engineering practice, is full of trade-offs. The ability to spot the important issue among many issues is crucial. However, it is not easy to learn how to ignore what is not important. This takes time.

In summary, working as an IC implementation engineer requires unique knowledge and skills.

Conclusion

The most important lesson that I have learned during my career as an ASIC chip integration engineer is this: it takes a solid understanding of key concepts, such as those addressed in this text, to stand out as a top-of-the-line engineer. I have seen engineers with many years' experience unable to compete against relatively new engineers. I have also seen the results of weak technical leadership: important projects delayed and, consequently, market opportunities lost. Moreover, I have seen engineers who, while mastering specific tools, cannot make bigger, more significant contributions to projects. All of these problems can be traced back to one source: a lack of correct understanding of the concepts.

On the other hand, engineers with a solid understandings of the concepts can:

- Learn new tools with great ease
- Work with greater efficiency, and, most importantly
- Avoid making mistakes

In most cases, avoiding mistakes is the best way to compete; it is the best way to achieve something before the competitors do. A good ASIC engineer should be a "problem-driven" worker, not "tool-driven" worker. In this information age, nobody has the capability of knowing every command of every tool, especially in the SoC integration arena where new tools are surfacing at unforeseen speed. Only by mastering the underlying basic concepts can he/she become the master of the tools, not be mastered by the tools.

Acronyms

ADC	Analog-to-Digital Converter
ARC	Argonaut RISC Core
ARM	Advanced RISC Machine
ATE	Automated Test Equipment
ATPG	Automatic Test Pattern Generation
ASIC	Application-Specific Integrated Circuit
BIST	Built-in Self-Test
CAD	Computer-Aided Design
CMOS	Complementary Metal Oxide Semiconductor
CPU	Central Processing Unit
CTS	Clock Tree Synthesis
DAC	Digital-to-Analog Converter
DDFT	Dynamically Deactivated Fast Transistors
DDR	Double Data Rate
DEF	Design Exchange Format
DFM	Design for Manufacturability
DFT	Design for Testability
DLL	Delay Lock Loop
DPPM	Defect Part Per Million
DRAM	Dynamic Random Access Memory
DRC	Design Rule Check
DSP	Digital Signal Processing
DSPF	Detailed Standard Parasitic Format
DV	Design Verification
ECO	Engineering Change Order
EDA	Electronic Design Automation
EDIF	Electronic Design Interchange Format
EEPROM	Electronically Erasable Programmable Read-Only Memory
EM	Electromigration
ERC	Electric Rule Check
ESD	Electrostatic Discharge

VLSI Circuit Design Methodology. By Liming Xiu

ESL	Electronic System Language
FPGA	Field Programmable Gate Array
GALS	Globally Asynchronous Locally Synchronous
GDSII	Gerber Data Stream Information Interchange
GOI	Gate Oxide Integrity
GsAs	Gallium Arsenide
GUI	Graphic User Interface
GV	Geometry Verification
HCS	Hot Carrier Stress
HDL	Hardware Description Language
IC	Integrated Circuit
I/O	Input/Output
IP	Intellectual Property
ITRS	International Technology Roadmap for Semiconductors
LVS	Layout Versus Schematic
nm	nanometer, 1×10^{-9} m
MCM	Multichip Module
MIPS	Microprocessor without Interlocked Pipeline Stages
MOSFET	Metal Oxide Semiconductor Field Effect Transistor
MTBF	Mean Time Between Failures
NBTI	Negative Bias Temperature Instability
NOC	Network on Chip
NRE	Nonrecurring Engineering
OCV	On-Chip Variation
OPC	Optical Proximity Correction
P&R	Place and Route
PCB	Printed Circuit Board
PG	Pattern Generation
PLL	Phase Lock Loop
PSM	Phase Shift Mask
PTV	Process, Temperature, Voltage
RADHARD	Radiation Hardened
RAM	Random Access Memory
RF	Radio Frequency
ROI	Return on Investment
ROM	Read-Only Memory
RTL	Register Transfer Level
SAF	Stuck at Fault
SDC	Standard Design Constraint
SDF	Standard Delay Format
SERDES	SERializer/DESerializer

SIP	System in Package
SOC	System on Chip
SOI	Silicon on Insulator
SPEF	Standard Parasitic Exchange Format
SPICE	Simulation Program with Integrated Circuit Emphasis
SRAM	Static Random Access Memory
SSST	Statically Selected Slow Transistors
STA	Static Timing Analysis
SV	Schematic Verification
TSP	Traveling Salesman Problem
μm	Micrometer, 1×10^{-6} m
USB	Universal Serial Bus
VDSM	Very Deep Submicron
VHDL	VHSIC Hardware Description Language
VHSIC	Very High-Speed Integrated Circuit
VLSI	Very Large Scale Integration

Bibliography

1. Y. Nishi and R. Doering, *Handbook of Semiconductor Manufacturing Technology,* New York: Marcel Dekker, 2000.
2. R. L. Geiger, P. E. Allen, and N. R. Strader, *VLSI Design Techniques for Analog and Digital Circuits,* New York: McGraw-Hill, 1990.
3. R. J. Baker, H. W. Li, and D. E. Boyce, *CMOS Circuit Design, Layout, and Simulation,* New York: IEEE Press, 1998.
4. M. J. S. Smith, *Application-Specific Integrated Circuit,* Reading, MA: Addison-Wesley, 1997.
5. J. F. Wakerly, *Digital Design, Principle and Practices,* Englewood Cliffs, NJ: Prentice-Hall, 1994.
6. A. S. Sedra and K. C. Smith, *Microelectronic Circuits,* 2nd ed., New York: Holt, Rinehart and Winston, 1987.
7. N. H. E. Weste and K. Eshraghian, *Principles of CMOS VLSI Design—A System Perspective,* Reading, MA: Addison-Wesley, 1992.
8. L. A. Glasser and D. Dobberpuhl, *The Design and Analysis of VLSI Circuits,* Reading, MA: Addison-Wesley, 1985.
9. D. A. Hodges, *Analysis and Design of Digital Integrated Circuits,* New York: McGraw-Hill, 2003.
10. J. Schroeter, *Surviving the ASIC Experience,* Englewood Cliffs, NJ: Prentice-Hall, 1992.
11. N. Sherwani, *Algorithms for VLSI Physical Design Automation,* 2nd ed., Norwell, MA: Kluwer Academic Publishers, 1995.
12. S. S. Sapatnekar and S. M. Kang, *Design Automation for Timing-Driven Layout Synthesis,* Norwell, MA: Kluwer Academic Publishers, 1993.
13. W. Wolf, *Modern VLSI Design: A Systems Approach,* Englewood Cliffs, NJ: Prentice-Hall, 1994.
14. L. Bening and H. Foster, *Principle of Verifiable RTL Design: A Functional Coding Style Supporting Verification Processes in Verilog,* Norwell, MA: Kluwer Academic Publishers, 2000.
15. P. Rashinkar, P. Paterson, and L. Singh, *System-on-a-Chip Verification,* Norwell, MA: Kluwer Academic Publishers, 2001.
16. H. Johnson and M. Graham, *High-Speed Digital Design: A Handbook of Black Magic,* Englewood Cliffs, NJ: Prentice-Hall PTR, 1993.
17. I. Sutherland, B. Sproull, and D. Harris, *Logical Effort, Designing Fast CMOS Circuits,* San Francisco: Morgan Kaufmann Publishers, 1999.

VLSI Circuit Design Methodology. By Liming Xiu
Copyright © 2008 the Institute of Electrical and Electronics Engineers, Inc.

18. P. R. Gray, P. J. Hurst, S. H. Lewis, and R. G. Meyer, *Analysis and Design of Analog Integrated Circuits,* 4th ed., New York: Wiley, 2001.

19. K. R. Laker and W. M. C. Sansen, *Design of Analog Integrated Circuits and Systems,* New York: McGraw-Hill, 1994.

20. P. J. Ashended, *The Designer's Guide to VHDL,* San Francisco: Morgan Kaufmann Publishers, 1996.

21. International Technology Roadmap for Semiconductors (2005), http://public.itrs.net.

22. P. R. Groeneveld, "Physical design challenges for billion transistor chips," in *Proceedings of IEEE International Conference on Computer Design,* pp. 78–83, 2002.

23. L. Xiu, "Several new concepts to bridge the 'logic effort' research and SoC timing closure practice," in *Proceedings of IEEE Dallas/CAS Workshop,* pp. 229–135, 2005.

24. R. Knoth, "Time is on my side—Understanding and winning with clocks," Presented at 2005 Magma Users Summit on Integrated Circuit, Santa Clara, CA, 2005.

25. B. Zahiri, "True DFM/DFY solutions require more than OPC and DRC," *Chip Design,* June/July, 31–34, 2006.

26. T. Pompl, "Practical aspects of reliability analysis for IC designs," in *Design Automatic Conference 2006,* pp. 193–198, San Francisco, CA.

27. C. F. Hawkins and J. Segura, "Test and reliability: Partners in IC manufacturing, Part 1," *IEEE Design & Test of Computer,* vol. 16, issue 3, pp. 64–71, 1999.

28. C. F. Hawkins and J. Segura, "Test and reliability: Partners in IC manufacturing, Part 2," *IEEE Design & Test of Computer,* vol. 16, issue 4, pp. 66–73, 1999.

29. S. Furber, *ARM System-on-Chip Architecture,* 2nd ed., Reading, MA: Addison-Wesley, 2000.

30. N. D. Liveris, H. Zho, and P. Banerjee, "An efficient system-level to RTL verification framework for computation-intensive applications," in *Proceedings of 14th Asian Test Symposium,* pp. 28–33, 2005.

31. *ASIC/IC Design-for-Test Process Guide,* Version 8.5_1, Mentor Graphics, 1991.

32. "Gain-based synthesis: Speeding RTL to silicon," Magma white paper, Available: http://www.magma-da.com.

33. G. E. Moore, "Cramming more components onto integrated circuits," *Electronics, 38,* 82–85, 1965.

34. K. Roy, S. Mukhopadhyay, and H. Mahmoodi, "Leakage current mechanisms and leakage reduction techniques in deep-submicrometer CMOS circuit," in *Proceedings of IEEE, 91,* 305–327, 2003.

35. D. Lee, D. Blaauw, and D. Sylvester, "Gate oxide leakage current analysis and reduction for VLSI circuits," *IEEE Trans. on VLSI, 12,* 155–166, 2004.

36. N. S. Kim, T. Austin, D. Blaauw, K. Flautrer, J. S. Hu, M. J. Irwin, M. Kandemir, and V. Narayanan, "Leakage current: Moore's law meets static current," *Computer, 36,* 68–75, 2003.

37. R. Sridhar, "Clocking and synchronization in sub-90nm system-on-chip (SoC) designs," in *Proceedings of IEEE Dallas/CAS Workshop,* pp. 49–84, September 2004.

38. B. Razavi, "CMOS technology characterization for analog and RF design," *JSSC, 34,* 268–276, 1999.

39. L. E. Larson, "Integrated circuit technology options for RFIC's: present status and future directions," *JSSC, 33,* 387–399, 1998.

40. B. Razavi, "Challenges and trends in RF design," in *Proceedings of 9th IEEE International ASIC Conference,* pp. 81–86, September 1996.

41. M. Kulkarni, A. Marshall, C. R. Cleavelin, X. Weize, C. Pacha, K. von Armin, T. Schulz, K. Schruefer, and P. Patruno, "Ring oscillator performance and parasitic extraction simulation in finfet technology," in *Proceedings of 2006 IEEE Dallas/CAS Workshop,* DCAS-06, pp. 123–126, 2006.

42. C. A. Mack, "The end of the semiconductor industry as we know it," plenary address, presented at *Optical Microlithography XVI,* Feb. 25–28, 2003.

43. C. A. Mack, "The new, new limits of optical lithography," Emerging Lithographic Technologies VIII, in *Microlithography 2004,* Feb. 22–27, pp. 1–8, 2004.

44. C. A. Mack, *Field Guide to Optical Lithography,* SPIE Field Guide Series, FG06, Bellingham, WA: The International Society for Optical Engineering (SPIE), 2006.

45. G. Riley, "Bump, dip, flip: Single chip," in *Proceedings of Surface Mount International,* pp. 535–541, 1997.

46. J. H. Lau (Ed.), *Flip Chip Technologies,* McGraw-Hill: New York, 1995.

47. C. Pixley, A. Chittor, F. Meyer, S. McMaster, and D. Benua, "Functional verification 2003: Technology, tool and methodology," in *Proceedings of 5th International Conference on ASIC,* pp. 1–5, October 21–24, 2003.

48. R. Wang, W. Zhan, G. Jiang, M. Gao, and S. Zhang, "Reuse issues in SoC verification platform," in *Proceedings of 8th International Conference on Computer-Supported Cooperative Work in Design,* pp. 685–688, 2003.

49. P. James, *Verification Plans: The Five-Day Verification Strategy for Modern Hardware Verification Languages,* Norwell, MA: Kluwer Academic Publishers, 2004.

50. E. G. Friedman, *Clock Distribution Networks in VLSI Circuits and Systems,* New York, IEEE Press, 1995.

51. E. G. Friedman, "Clock distribution networks in synchronous digital integrated circuits," *Proc. IEEE, 89,* 665–692, 2001.

52. S. C. Chan, K. L. Shepard, and P. J. Restle, "Design of resonant global clock distributions," in *Proceedings of IEEE International Conference on Computer Design,* pp. 248–253, 2003.

53. J.-Y. Chueh, M. C. Papaefthymiou, and C. H. Ziesler, "Two-phase resonant clock distribution," in *Proceedings of IEEE Computer Society Annual Symposium on VLSI,* pp. 65–70, 2005.

54. B. El-Kareh, A. Ghatalia, and A. V. S. Satya, "Yield management in microelectronic manufacturing," in *Proceedings of Electronic Components and Technology Conference,* pp. 58–63, 1995.

55. S. P. Cunningham, C. J. Spanos, and K. Voros, "Semiconductor yield improvement: results and best practices," *IEEE Trans. Semicond. Manuf., 8,* 103–109, 1995.

56. C. Zhou, R. Ross, C.Vickery, B. Metteer, S. Gross, and D. Verret, "Yield prediction using critical area analysis with inline defect data," in *Proceedings of IEEE Advanced Semiconductor Manufacturing Conference and Workshop,* pp. 82–86, 2002.

Index